# Securing a Connected Future

William Harding • Andreas Hartmann
Shayla O'Brien • Viktor Sinzig
Gabriel Break • Natalie Sinzig

# Securing a Connected Future

## IoT Cybersecurity & IP Security for Makers and Users of Medical Technology

Springer

William Harding
College of Graduate and
Professional Studies
Trine University
Ruskin, FL, USA

Shayla O'Brien
Product Analysis Lab (PAL)
Medtronic
Phoenix, AZ, USA

Gabriel Break
Product Sustainability
Medtronic
Tel Aviv, Israel

Andreas Hartmann
Manufacturing Systems & AI
Digital Manufacturing Ireland (DMI)
Craughwell Co, Galway, Ireland

Viktor Sinzig
Global Customer Adoption and
Organizational Change Management
PTC
San Francisco, CA, USA

Natalie Sinzig
San Francisco, CA, USA

ISBN 978-3-032-07308-2 ISBN 978-3-032-07309-9 (eBook)
https://doi.org/10.1007/978-3-032-07309-9



This Springer imprint is published by the registered company Springer Nature Switzerland AG
The registered company address is: Gewerbestrasse 11, 6330 Cham, Switzerland

# Purpose of This Book

Currently, very little relevant material written by experts on the Internet of Things (IoT), cybersecurity, and Intellectual Property (IP) security industry is tailored specifically for the creators, consumers, and protectors of medical technology. This book will address this gap by demonstrating how robust cybersecurity measures are essential for protecting intellectual property and IoT-connected medical devices. By intertwining these two critical areas, we aim to provide a unified framework that enhances the security and value of medical technology innovations. With those points in mind, the authors of this book will explore the topics of cybersecurity and IP within the scope of the IoT lifecycle, emphasizing medical technology, and providing readers with the knowledge to recognize vulnerabilities at each stage of the IoT and IP development lifecycle. This book will enable organizations to recognize risks and threats as they proactively incorporate robust security measures into their design, development, and protection processes. Accordingly, the topics discussed in this book will be (1) protecting sensitive data as it is created, transformed, transmitted, and aggregated, (2) mitigating potential attacks, (3) the impacts of third-party acquisitions on security, (4) case studies that emphasize potential risks, (5) pros and cons of using AI, (6) identification and management of vulnerabilities, and finally (7) examining emerging trends and opportunities associated with integrating new technologies, with an eye toward the future of cybersecurity.

# Preface

The six authors of this book are colleagues and friends with many combined years (~125) in the design, development, and manufacturing of medical devices, as well as in the development and defense of intellectual property (IP). It is from those years in industry that the team of authors recognized and sought to promote the application of advanced cybersecurity methods in the protection of every aspect of a medical device (i.e., from idea to decommissioning). Moreover, the team has not only researched and written on the topics of cybersecurity, IoT, and IP, but they have also worked as developers and integrators in the industry, building secure systems, leading programs, and dealing with the day-to-day challenges of applying cybersecurity practices within organizations. It was during that journey of practical cybersecurity application that the team realized that no one had created material that recognized that protecting data, processes, hardware, and software was as important as protecting ideas. Thus, the team was stimulated to create a book that would help future engineering and legal teams work together toward a shared goal of protecting ideas and solutions that ultimately improve a patient's quality of life.

The effort of writing this book was not an easy one, and without the support of Springer Nature, this book may not have existed. But what helped the team succeed in writing this book was that each author is highly skilled in time management and project development, and possesses an amazing ability to communicate as well as collaborate. As a result of this ability to collaborate, the team easily defined roles, established a regular cadence of meetings, effectively managed objectives, and merged material into a book that speaks with one voice as it seeks to inform readers. Similarly, the team shares a vision of what the future of medical technology might look like, and that this book will stimulate inventors, engineers, and leaders to be better prepared as they implement protective measures. That said, the team knows that there is a need to improve awareness of cybersecurity risks associated with the invention and manufacturing of medical devices, which is why this book was created, aiming to be a guide for industry leaders who make decisions in the manufacturing of secure medical solutions.

Ruskin, FL, USA William Harding
Craughwell Co, Galway, Ireland Andreas Hartmann
Phoenix, AZ, USA Shayla O'Brien
San Francisco, CA, USA Viktor Sinzig
Tel Aviv, Israel Gabriel Break
San Francisco, CA, USA Natalie Sinzig

# Contents

# Understanding the IoT Lifecycle and Associated Cybersecurity Risks

## 1 What Is an IoT Lifecycle

The lifecycle of IoT should not be considered as a linear process, where instead it embodies numerous phases, where each or several phases can be envisioned as iterative loops. Specifically, according to Harding (2022), those phases are technology exploration, decision-making, design, development, adoption, testing, integration, sustainability, and end-of-life, which represent the primary phases associated with an IoT solution's lifecycle. Those phases, viewed through the lens of managing the IoT lifecycle, are the conception of an idea, design, manufacturing, deployment, and then decommissioning/archival of the IoT solution (Fig. 1). Furthermore, each phase may engage different players/actors as the IoT solution moves through its lifecycle. Those players could be innovators, solution sponsors, Information Technology (IT) specialists, engineers, quality assurance specialists, regulatory agents, technicians, production associates, and even end customers, such as patients, caregivers, clinical technicians, and healthcare professionals. Accordingly, each player represents a relevant stakeholder, with skin in the game.

In addition to the principal phases, there are many nuances associated with those phases of the IoT lifecycle, including:

- <u>Solution Definition:</u> Represents a coherent approach to discussing the scope and specifics of the solution to be implemented.
- <u>Requirements Gathering:</u> Collecting, documenting, and understanding the needs, expectations, and boundaries of users/customers to ensure that a project, system, or solution is developed to meet their objectives.
- <u>Hardware/Software Design</u>: The creation of detailed plans and specifications for both the hardware and software elements of a system to ensure that they work together in meeting the desired functionality, performance, and user requirements.



W. Harding et al., *Securing a Connected Future*,
https://doi.org/10.1007/978-3-032-07309-9_1

**Fig. 1** Technology development lifecycle (Wikimedia Commons)

- Network/Communications Architecture: To design and plan the structure, protocols, and components required for the effective and efficient communication and transformation of data within a network between devices, systems, or users.
- Component Selection: The process of choosing the right hardware, software, or system components to ensure that they reliably and effectively meet project, system, and product requirements.
- Solution Prototyping: Promotes the early creation and simplification/modeling of a product, system, or solution that enables stakeholders, designers, and developers to evaluate, demonstrate, and refine concepts before developing a full-scale solution.
- Solution Development: The phase/process where solution developers create a product that complies with the solution requirements.
- Integration with Existing Systems: Ensures that new solutions work smoothly with the intended target infrastructure, applications, humans, or technologies that are already in place within the organization.
- Solution Testing: Helps ensure that a developed and finalized solution meets the requirement specifications, operates as expected, and satisfies the needs of the stakeholders.
- Quality Assurance: Ensures that a solution follows standards and consistently meets the highest quality requirements throughout a solution's development lifecycle.
- Regulatory Compliance: Ensures that a market-ready solution complies with the laws, regulations, standards, and guidelines set by governing bodies, industry authorities, and legal frameworks.
- Solution Refinement/Improvement: Promotes the enhancement and optimization of a solution after initial development and deployment based on stakeholder feedback, solution performance data, and solution monitoring.
- Solution Configuration: Proposes the modification of a solution, such that it meets specific requirements, customer/user needs, or other stakeholder preferences without altering the solution's core structure or underlying code (i.e., in the case of software).

- User Training: Equips users with the knowledge and skills needed, such that they can effectively and efficiently use a solution.
- Solution Performance/Security Monitoring: Promotes the continuous tracking and assessment of a solution to ensure the health, efficiency, and security of the deployed solution.
- Solution Maintenance: Ensures that a deployed solution remains operational, up-to-date, secure, and aligned with gradually changing business and customer needs.
- Decommissioning/Disposal of Sensitive Data: Ensures the secure and responsible elimination of data that is no longer needed, while making sure that the data cannot be recovered or misused.

The principal lifecycle phases might appear on the surface as being fully inclusive of the nuances presented, but if overlooked or ignored, failure to consider these details will result in a solution that doesn't meet user and customer expectations.

## 2 The Importance of IoT Medical Security

Within the scope of IoT systems associated with the healthcare/medical industry, it is important not to myopically consider only fixed-mounted devices, such as repeaters, extenders, routers, hubs, access points, switches, sensors, etc. The scope should also include wearables, implantables, consumables (e.g., PillCam), vehicles/transportation, BYoD (i.e., user-supplied mobile devices), security devices, environmental monitoring, and communications devices. Respectively, the list can be exhaustive with considerable room to grow, and the need for security is paramount. The exchange of data between IoT-enabled devices requires a perspective that should stimulate solution architects to examine the exchange between IoT tech as literally the transformation of data. This stimulates the need to ensure the integrity of the data aligns with the need to protect the data (Kor et al. 2023).

Of course, there are many risks associated with IoT tech, but the lack of awareness associated with securing IoT data transfers and transformations could manifest as a loss of patient data, compromise of protected intellectual property (IP), reduced market share, reduced patient quality of care, and potentially loss of life. Additionally, less-than-obvious risks could be realized in the scientific and academic research environments, resulting in inconclusive/inaccurate outcomes due to the use of compromised/corrupted data. So, without diluting, diminishing, or overwhelming designers and users of IoT tech, the need for IoT security must always be kept in mind beyond the initial deployment of the tech and throughout the lifecycle of IoT.

Thinking in terms of personal devices (e.g., BYoD), it is known that healthcare systems/devices aggregate many digital technologies. And while the use of numerous technologies might be good for patients, it also means that patients could be the target of cyber threats. Compared to other industries, healthcare is more prone to cyberattacks (Garcia-Perez et al. 2023). This is mainly because the security of healthcare systems is not robust, and the complexity of the digital ecosystem makes it much more difficult to maintain and defend. That point is especially true with end-of-life medical devices, where the companies do not have the resources or funding to keep up with potential cybersecurity threats.

To get a better understanding of how big the issues with healthcare technology security are, a 2017 survey indicated that healthcare was one of the most attacked sectors worldwide (Jarrett 2017). In fact, 81% of the 223 organizations surveyed had been attacked. In the US alone, more than 110 million patients had their data stolen in 2015. And with cybersecurity attacks on the rise, recent research reveals that in 2024, 88% of healthcare workers and 90% of healthcare organizations have experienced phishing attacks (Adesokan-Imran et al. 2025).

So, what kind of security dangers exist, and how bad can they be? First, it is important to understand the depth of cybersecurity risks and the fact that healthcare systems are likely targets, where the issue becomes even more challenging because many hospitals are slow to upgrade their current equipment, particularly their older systems. This reluctance to change makes hospitals an easy target for cyber-attacks.

Take, for instance South-eastern Norway regional health authority. In 2018, Protected Health Information (PHI) had records of nearly 2.9 million people compromised, where the attack came from a vulnerability in a Windows XP legacy system. Luckily, there were no apparent risks to patient safety or hospital operations, but the event highlighted the importance of the expectations captured in the General Data Protection Regulation (GDPR). The GDPR is the European Union's privacy law that not only protects medical information like HIPAA but also encompasses all personal data.

When thinking of IoT in medical devices, Personal Medical Devices (PMDs) first come to mind. Those PMDs can be implanted in the patient's body or attached to the patient (e.g., wearable health devices, fitness devices, etc.). And some of those PMDs might not be required to follow security requirements standards or government regulations, where they might represent an easy target for cybersecurity attacks, for the following reasons:

- Limited encryption due to the reduced inclusion of security elements that would overburden battery power consumption and the use of poorly managed transmission practices, which may lead to confidentiality, integrity, availability, and privacy exposure.
- Authentication for wireless communication. This makes the data stored on the PMDs at risk of being disclosed to unauthorized individuals as well as to malicious updates, which could represent greater security concerns.

The importance of the risks associated with PMDs are embedded in the concerns that many individuals and groups have expressed when examining all the lifecycle phases of PMDs (Harding 2022). For example, data and device security experts state that cybersecurity has only been considered during the design phase of the medical devices, but considering the risks involved, these cybersecurity issues should be addressed throughout the entire lifespan of the PMDs. This ensures that patient safety and data integrity remain uncompromised.

## 3 Identifying Potential Product and Data Cybersecurity Risks

When thinking in terms of an IoT solution's lifecycle and aligning that process with the application of cybersecurity, many individuals/groups believe that cybersecurity should only be considered at the integration phase or possibly at the adoption phase. However, it is proposed that IoT cybersecurity must be considered at the conception of an idea. Respectively, that point becomes clear as the innovator considers everything down to the discrete components and subassemblies that will be used to create their IoT solution. For example, cybersecurity risks must be kept in mind as one selects high-quality components to populate their subassemblies. Without that thought process, imagine incorporating a component that is substandard, which might result in a circuit failure, radiation of data using unshielded components, or allowing a backdoor/vulnerability where malicious attackers can compromise the IoT solution. That said, cybersecurity must be at the forefront when developing IoT solutions and remain at the forefront through the entire lifecycle of those solutions (Barrera et al. 2023).

Thinking now about how to reduce risks within an IoT solution's lifecycle, there are several known points to consider, as captured in the following list:

- Quality discrete components
- Tested and validated subassemblies
- Vendor and in-house-developed components/systems
- Accidental exposure of protected data
- Malicious attacks (e.g., brute force, remote, denial of service, etc.)

As can be imagined, discrete components are a potential point of failure, specifically as it relates to being the weak link in the chain, but ensuring that components are validated against industry standards, one can reduce that risk. That point also emphasizes the need to follow industry standards associated with placing and securing discrete components on a circuit board (i.e., PWB/PCB). That said, imagine an antenna integrated into a system that allows a signal to interfere with the primary communications function of an implanted medical device. For example, a damaged doctor's device could transmit a signal that could damage a patient's device. Thus, we need to guard against unintentionally compromising a patient's medical device.

But what about subassemblies? Are they not safe to use in a medical device? Well, if you are not familiar with the term subassemblies, those are an aggregation of discrete components that are assembled to perform a specific function and to be consumed into a final assembly. For example, a Bluetooth module or an RF receiver might contain a subassembly that is consumed into a device or system that requires communications over WiFi or Bluetooth Low Energy (BLE). And those subassemblies might be built by another internal group or a third-party vendor/supplier.

Moving back to the original questions regarding if subassemblies are safe to use in a medical device, the answer is not a straightforward one. Because if subassemblies do not use reliable components, are not validated against the medical device's requirements, and if those subassemblies are not tested and not compliant with cybersecurity standards (e.g., IEEE/UL 2933-2024), then can you trust them with your data and your life? Additionally, you need to consider who is doing the testing and validation of those subassemblies. Is it the consumer, third-party supplier, component manufacturer, device manufacturer, or the doctor? Well, the answer is that everyone should test any product that they receive. When it comes to protecting humans and their data, trust no one without seeing evidence of compliance and the following of regulated standards (i.e., zero-trust).

With those points in mind, just as you would check an electrical device for the UL or CE mark to make sure that it was safe, you should do the same for any device that would be used to monitor vitals, diagnose health states, and administer therapy. And that does not just apply to the final product. It should apply to discrete components and subassemblies. Because, though a subassembly might appear to meet your Bluetooth needs, how do you know that it was meant to be used with a pacemaker, or how do you know that it would not interfere with other subassemblies or even other devices? You do not, and that is why it is important that testing be completed at the unit level and at the system level, where the integration and testing of components and subassemblies must produce evidence of compliance, functionality, durability, and safety.

Turning to an examination of the risks associated with the entire list of IoT lifecycle phases, steps, and processes, the following expanded explanation must be considered before moving to the next phase, step, or process:

## 4 Risks Associated with IoT Lifecycle Phases

### *4.1 Lack of Security Requirements—Requirements Gathering Phase*

Lack of attention to detail during the requirements-gathering phase means that all needed security requirements are not integrated into a solution's design. As a result, a solution may be vulnerable to exploits and attacks because foundational security principles were not considered during requirements gathering.

### *4.2 Inadequate Threat Modeling—System Design Phase*

During the system design phase, failing to properly identify potential threats and vulnerabilities in the system can leave the solution exposed to cyberattacks later. Additionally, not creating an appropriate threat model may result in a need to perform costly rework to correct for not considering potential security flaws and undetected vulnerabilities.

## 4.3 Use of Insecure Technologies—Hardware/Software Design Phase

Decision-making technology stacks, platforms, and third-party components may not consider known vulnerabilities. The result of using unsecured and/or outdated software/hardware may introduce risks, such as exploits, data breaches, or system failures. Furthermore, if solution prototypes are built too quickly, such that concepts are not fully tested and properly reviewed by the security team, then design flaws might allow vulnerabilities to be exploited later. For example, improper input/output validation, unsecured APIs, and unsafe data handling may put the solution users at risk.

## 4.4 Integration of Security Controls—Integration with Existing Systems Phase

There is a risk that security controls (e.g., trust, identity, privacy, protection, safety, and security) may be incorrectly defined or omitted from the phase of integrating a solution with existing systems. As a result of incorrectly defining security controls, sensitive data may be exposed, which increases the potential for increased cybersecurity attacks. For example, solutions often integrate with existing and new systems (e.g., third-party APIs and databases), which represent unique security issues, such as backend vulnerabilities, outdated APIs, and unsecured integration points that can expose solutions to attackers.

## 4.5 Insufficient Data Protection Considerations—Data Definition Phase

As is the case when developing a solution that consumes, stores, transforms, or transmits data, there is a need to define the relevant data. Further, during the definition of data phase, one must consider how data will be protected. However, if proper data security measures are not considered, sensitive data might not be adequately encrypted or securely stored, leading to privacy violations, regulatory non-compliance, or data breaches. For example, solution prototyping often uses real data or simulated data that mimics actual production environments, which exposes sensitive and proprietary information due to a lack of proper encryption or data handling practices. Some of that simulated data might be stored as plaintext passwords or other sensitive data transmitted over unencrypted channels.

In addition to sensitive data being exposed during solution prototyping, QA teams also use the same data for real-world testing of production environments, which can include personal information, payment details, and intellectual property

(IP). Without proper data protection, leaks and unauthorized access can occur, which also might represent non-compliance with privacy regulations (e.g., GDPR or CCPA) that can lead to severe penalties and damage to an organization's reputation. Accordingly, it is suggested that encrypted, anonymized, and/or synthetic data should be used during the prototyping and testing phases.

## *4.6 Lack of Compliance and Regulatory Awareness—Regulatory Assessment Phase*

If (during the regulatory assessment phase) there is a lack of awareness or an inability to ensure compliance with regulatory requirements, the potential is increased that the solution may not account for relevant industry practices, regulations, and security standards (e.g., HIPAA, GDPR, etc.). Additionally, non-compliance with regulations can result in legal and financial penalties, as well as damage to the organization's reputation.

## *4.7 Unclear Security Roles and Responsibilities and Lack of Training—Role Definition Phase*

Within the scope of relevant stakeholders, it is important to define roles and responsibilities associated with applying security. If security roles across a solution project are vague and/or not clearly defined, there exists the potential for security tasks to be overlooked or delayed, making it harder to enforce security throughout the project lifecycle. Additionally, if the role of the user/customer is overlooked or not considered (i.e., how users interact with the solution and appropriate user security training), then there exists a potential for increased human error related to phishing attacks, breaches, and compromised weak passwords.

## *4.8 Overlooking Secure Development Practices—Development Phase*

If development practices, such as secure coding standards, code reviews, and automated security testing, are not prioritized during the solution definition phase, it can lead to vulnerabilities (e.g., SQL injection, buffer overflows, etc.) in the software, which can later be exploited. For example, if security best practices (e.g., following standards, using secure development libraries, and applying proper encryption) are skipped during the prototyping phase, then sensitive data might be exposed, resulting in cyber-attacks.

## 4.9 Third-Party Vendor Risk Management—Planning, Libraries. Dependencies, Tools, and Supply Chain Phase

Medical devices and solutions often use third-party elements, such as subassemblies, APIs, tools, or services, and if those elements do not undergo rigorous security evaluations, then the final solution could be exposed to breaches. Further, if third-party vendors do not follow security standards (e.g., IEEE/UL 2933-2024) and use open-source tools/libraries, the lack of compliance and validation might introduce vulnerabilities, resulting in exploitation.

In addition to the risk associated with integrating unvalidated third-party elements, there are also risks associated with using third-party tools such as Quality Assurance (QA) during the testing and validation of solutions. For example, unvalidated third-party tools might inadvertently introduce malware or expose the solution to other security gaps. To mitigate these risks, third-party elements need to be regularly reviewed, tested (e.g., OWASP Dependency-Check), and updated to ensure that standards are being followed and that trusted sources are being used.

## 4.10 Misconfiguration/Misalignment Risks—Solution Configuration Phase

If a solution is defined with assumptions about the configuration of security controls (e.g., firewalls, permissions, etc.) without testing or validating those assumptions, the poorly configured system can leave it exposed to unauthorized access or other security threats, specifically in the network or application layer. That point is even more relevant when considering a system that is poorly configured, might also not be developed with business continuity in mind (e.g., disaster recovery, event response, etc.). Whereas a lack of planning for business continuity can leave an organization vulnerable to service disruptions in the event of a cyberattack (e.g., ransomware) or data loss.

## 4.11 Inadequate Testing and Validation—Solution Testing Phase

During the solution testing phase, many risks must be considered. Some of those risks are listed below, followed by the implications associated with those points:

**Testing and Validation**

**Risk** The testing and validation phase may not include provisions for security testing and validation (e.g., penetration testing, vulnerability assessments) to identify weaknesses early.

**Result**: Security flaws might be detected later, requiring significant rework and delaying the overall project.

**Prototypes**

**Risk** Prototypes are often tested in non-production environments that may not be adequately secured, leading to misconfigurations that could allow unauthorized access or data leakage. For example, exposed test databases, insecure default settings, or failure to restrict access to the prototype.
**Result** Test environments may be used as a backdoor into the production system or leak valuable information about the project's design.

**Security Testing**

**Risk** In a rush to iterate and validate the solution's functionality, security testing might be insufficient, leaving the system open to common attack vectors like SQL injection, cross-site scripting (XSS), or cross-site request forgery (CSRF). For example, inadequate penetration testing, fuzz testing, or vulnerability scanning. Additionally, QA processes sometimes focus primarily on functional testing, neglecting security testing, such as penetration testing, vulnerability scanning, or static/dynamic code analysis. This can lead to critical vulnerabilities going undetected until production.
**Result** Vulnerabilities can be exploited to attack the prototype, compromising its integrity or leading to security breaches.

## 4.12 *Lack of Proper Authentication and Authorization Controls—Prototyping Phase*

Prototypes are often developed with minimal authentication and authorization mechanisms, or these mechanisms may be incomplete, leaving the solution vulnerable. For example, weak or absent authentication (e.g., no multifactor authentication), inadequate role-based access control (RBAC), or open API endpoints. Additionally, if prototypes are deployed in public or untrusted environments (e.g., cloud infrastructure or public-facing networks), they may inadvertently expose security weaknesses (e.g., improper firewalls, access controls, or monitoring). The results of those oversights may result in attackers gaining unauthorized access, escalating privileges, or bypassing security controls altogether. Attackers could exploit these exposures to breach the prototype or use those exposures to attack other systems.

Also, during the prototyping phase, if prototypes have limited logging and monitoring capabilities, then it can be difficult to detect or respond to security incidents during development and testing. For example, not logging key security events (e.g., failed login attempts, API misuse, etc.), or not monitoring traffic and system behaviors. Thus, without proper logs, identifying or mitigating attacks in real-time is difficult, leaving the prototype vulnerable to undetected exploitation. Furthermore, if developers have granted excessive permissions or broad access to systems and data during the prototyping phase, it could lead to unintentional exposure or misuse. For example, a case might exist when developers use production data in a development environment or use overly permissive roles in cloud services. This could result in exposing sensitive data or making security-critical changes without proper oversight.

When evaluating the role of the Quality Assurance (QA) processes in a new project, numerous cybersecurity risks need to be considered. For example, QA aims to ensure the compliance, functionality, security, performance, and quality of the project and the ultimate solution. However, it is possible that cybersecurity risks can be accidentally introduced during the QA validation and testing phases. Some of these risks and potential actions/mitigation measures are as follows:

**Malicious Code**

**Risk**: QA might accidentally introduce vulnerabilities into a solution during testing. Further, solution testers may not fully follow secure coding standards (e.g., by leaving debugging tools or hard-coded credentials in the codebase).
**Action**: Ensure that security reviews are part of the QA process and use automated tools to scan for hard-coded secrets, debug code, or other security risks.

**Incident Response**

**Risk**: During QA testing, vulnerabilities and data breaches might be discovered, and if there is an inadequate response plan in place, it might lead to slow detection and resolution.
**Action** The QA team must develop an incident response plan and ensure that QA resources are properly trained to recognize and respond to security threats.

**False Sense of Security**

**Risk**: During the QA testing phase, if a QA team makes the mistake of thinking that a system is secure when their test passes, it can lead to a false sense of security because QA testing generally focuses on functionality and does not always cover all potential attack vectors.
**Action**: To combat a false sense of security, the QA team must collaborate with security experts to verify all potential vulnerabilities. In conjunction with functional testing, the QA team must complete thorough security assessments and audits.

**Lack of Security Awareness**

**Risk**: There exists a risk that QA representatives have not received proper cybersecurity training, which can lead to human errors (e.g., leaving security vulnerabilities undetected, not following security best practices, or not identifying potential security flaws, etc.).

**Action**: Institute regular security training for all QA resources and associated stakeholders while promoting security awareness that ensures that QA resources understand secure coding principles, threat modeling, and attack vectors. Note: Outside QA resources who are trained in cybersecurity can be an option if internal QA resources are unavailable or lack proper cybersecurity training.

**Logging and Monitoring**

**Risk**: Because of QA's involvement in solution testing and compliance, QA's environment must be properly monitored and log usage. That logging must capture unauthorized access, attempts to exploit vulnerabilities, and/or system misconfigurations.
**Action**: Create a robust logging and monitoring system within the QA environment and ensure logs are collected, reviewed, and stored securely.

So, to wrap up the discussion of cybersecurity risks, always remember 'zero trust' unless evidence can be provided that gives you reason to trust what you receive from someone else. But why stop with physical components? Shouldn't the rule of zero-trust also apply to data? The answer is 'yes'. And that is where you can use the same rules about testing, validation, authentication, safety, security, identity checking, privacy confirmation, etc. You should expect the same level of reliability and safety for your data, or data that you consume, as you do for a medical device. In other words, the protection and safety of your device should be equally as important as the protection and safety of your data.

## 5 Unique Cybersecurity Issues Posed by Medical Devices (Fig. 2)

### *5.1 Connectivity & Data Transportation*

Imagine a medical device as a house on a deserted island, and the deserted island is like an isolated server that you might find within a hospital (i.e., safe from thieves/hackers). Then, over time, new people move onto the island, and they build their own houses. And since everyone on the island is a friendly neighbor, they start communicating with each other, kind of like medical devices might exchange information within the safety of the hospital server environment (i.e., across a local area network [LAN]). And since the local network server is isolated from the rest of the world, only local homes can communicate with each other. Everything is secure until a thief/hacker moves onto the island, secretly builds a bridge to the rest of the world, sending private conversations/data of the island neighbors off the island. And

**Fig. 2** Medical device hacking (Freepik.Com)

since the island neighbors assumed their data would not be shared, they never put in place any security measures (similar to a business firewall used to protect a server). Ultimately, as this scenario unfolds, the island neighbors have no idea that their homes are secretly being broken into and that off-island thieves/hackers might have now entered their homes and sabotaged them, leaving homeowners in a place where they are no longer safe to live.

Thinking about the island analogy, we are illustrating that we can never let our guard down when it comes to protecting our home or our medical devices. Similarly, with us not being in control of who moves onto the island, we are unaware of what devices might be added to a hospital network that may be compromised. Furthermore, the newly added devices could originate from a third party (i.e., a hospital partner or vendor), such as new robotic surgical technology or other technology that is meant to improve device data transfers. With that point in mind, devices should undergo a proper risk assessment prior to being incorporated into a hospital server to mitigate security vulnerabilities. Tied back to the island analogy, the homeowners need to assess their homes with security in mind, such as checking for proper locks on doors and windows.

## 5.2 Medical Cyber-Physical Systems (MCPSs)

Within hospitals, there are many different diagnostic, monitoring, or delivery devices, and some of those devices might be used to measure temperature, pulse rate, or even administer patient medicine. The hospital devices are part of what we call a mesh system, often associated with IoT constructs, and the devices can exchange and transform data interoperably. So, for example, if the pulse rate monitoring device detects that a patient's heart rate is exceeding normal limits, another device might respond by administering drugs that will help restore the patient's heart rate to a normal level. In short, we call the entire monitoring/diagnostic/delivery system a Medical Cyber-Physical System (MCPS). Given the purpose of MCPS, it is critical that they remain safe while ensuring that all patient data remains private and protected. Additionally, as presented in the island analogy, no strangers, thieves, or third-party vendors should have access to homeowner/patient data. Similarly, we should align that need for security with the need to ensure that devices are reliable and that healthcare professionals can trust the devices to securely perform the required function (Carreras Guzman et al. 2020).

## 5.3 Healthcare IT Infrastructure

Building off the previous 2 concepts and the island analogy, we can envision that the transfer and transformation of patient data is similar to vehicles safely transporting patients/data along streets, roads, and highways. And like a city's traffic infrastructure, the Vital Sync™ system (used to monitor many patients) is like a traffic control center, always watching all devices and the movement of patients/data. Correspondingly, just as the highway promotes the quickest movement of vehicles between 2 points, Vital Sync ™ uses a 2.4 GHz frequency as a method for quickly moving data across many devices. However, with the fast movement of data, some potential problems or risks need to be considered. So, to help understand the potential issues in more relatable terms, we offer the following 4 topics as examples worth consideration:

- Interference: Concerning interference that might negatively impact technology, consider the point that the previously discussed Vital Sync ™ device communicates over 2.4 GHz. Now consider that most home phones, many remote-controlled toys, etc., also communicate over 2.4 GHz, where using them in proximity to a Vital Sync™ device might interfere with the Vital Sync's™ critical function. That point emphasizes the need to ensure that critical healthcare devices can shield themselves from or adapt to interference.
- Network Congestion: Network Congestion might manifest in several different ways, like rush hour traffic on a highway, traffic jams can halt the movement of cars or, at a minimum, slow traffic below the normal speed limit. Similarly, if too many devices are running on a network, then the amount of data transfers might

be reduced and ultimately impact the speed at which critical devices need to perform. That point emphasizes the need for network monitoring tools and methods for giving critical systems higher priorities over less important systems.
- Co-channel Interference: With respect to co-channel interference, we compare the potential problems to the example of local neighborhood traffic using the same roads as highway traffic. And because we did not create different roads for highway traffic and local neighborhood traffic, we see how congestion and safety might become an issue. That example is similar to a hospital using the same network channel for both business and patient monitoring data. The problem of co-channel interference could be solved by using different channels for different functions.
- Adjacent-Channel Interference: Another form of interference that might represent a risk to medical device functionality is what we call adjacent-channel interference. And in the case of this last interference type, we use the analogy of having local traffic and highway traffic not being separated. The risk in that analogy is when we consider the need for high-speed traffic to get on/off from the highway (i.e., crossing into slow-moving local traffic) or if slower-moving local traffic inadvertently swerves into highway traffic. Similarly, if a hospital system uses the same frequencies for critical and business data or if the hospital network does not create barriers that divide data channels, then the issues of adjacent-channel interference will emerge.

Lastly, as pseudo-traffic police, hospital IT departments must find ways/methods to monitor and identify problems/risks associated with the movement of critical medical device data. Those processes/methods might mean setting up rules for when and how devices can use various communications frequencies and channels, or they might mean creating new methods (or highways) for devices to transmit/receive data (Williams and Woodward 2015; Dey et al. 2018).

## 6 Impact of Cybersecurity Breaches on Medical Device Products—Case Studies

There have been several examples over the years of medical devices being vulnerable to hacking. Since 2019, an estimated 25 billion dollars have been lost and 113,491 sensitive patient files have been lost due to cybersecurity attacks (Ordonez 2020). Many existing medical devices operate using older software, with minimal security, and which could put other devices at risk (Meggitt 2018).

One example was when a company notified patients that one of their insulin pumps was vulnerable to cyber hacking. A security consultant had discovered that the pump used an unencrypted radio signal, and therefore, a hacker in proximity could program the device. These cyberattacks are not only a data security risk but, even more importantly, a patient safety risk (Pycroft and Aziz 2018).

In 2017, the U.S. Food and Drug Administration (FDA) cited a medical device company for cybersecurity vulnerabilities within its pacemakers. Luckily, no devices were found to be tampered with, but the vulnerability was that a hacker could modify the pacing program, resulting in inappropriate pacing and/or premature battery depletion. Both incidents are a threat to patient safety. The vulnerability was corrected by deploying a software security patch, but patients had to go to their doctor's office to receive the update. It was noted to be a quick 3-min fix, but for some patients, it is cumbersome to get to a doctor's office, especially those living in rural areas. It would be better if the patch could have been deployed instantaneously across all devices, and the at-home monitoring systems were utilized to capture any abnormalities in pacing to ensure all patients receive the update. Tying this back to the lifecycle, devices need to be designed to handle updates with confidence so that patients can receive the needed software patch outside of a provider's office.

There have been multiple hospital attacks where hackers used medical devices (diagnostic equipment, therapeutic equipment, life support equipment, etc.) to move through the hospital network. It was discovered that the hackers were not only gathering confidential hospital data, but once the backdoor was established, the hackers had the potential to distort or change internal data. In one attack in Europe, three blood gas analyzers were affected with a backdoor established in the network. This allowed access to patient data, as well as made it possible to adjust gas analyzer parameters, causing direct harm to the patient (Ordonez 2020). One of the first hospitals to suffer a ransomware attack on electronic medical records was done so by the ransomware, which gained access to the system through a vending machine. Hospital staff were locked out of the system for a week, and a $17,000 ransom was paid. This further stresses the need for appropriate security measures to be put in place, including separating the connections of medical devices and other machines across a network.

## 7 Conclusion

As we've explored in this chapter, the IoT lifecycle introduces a complex web of cybersecurity risks that span from device design to decommissioning. For makers and users of medical technology, understanding these risks is only part of the equation. Similarly, it is important to recognize how medical device vulnerabilities intersect with the legal and strategic dimensions of intellectual property (IP). And as we move from exploring the phases of IoT technology, in the next chapter, we shift our focus to the legal frameworks and cybersecurity strategies that expose tools, methods, and processes that should be used to protect the enterprise value embedded within IoT-enabled medical devices. Further, by examining the convergence of IP law and cybersecurity, especially in the high-stakes realm of medical technology, we will uncover how organizations can safeguard innovative medical devices as well as proprietary knowledge and competitive advantage.

# Cybersecurity Related to IP and Protection of Enterprise Value

## 1 Intellectual Property (IP), Threats, and Protection

Filing patents and maintaining trade secrets are two popular ways that companies can protect their intellectual property (IP) rights. While patents are governed purely by federal law, trade secrets can be protected by both state and federal laws, which can vary both between state and federal protection schemes and depending on the state or jurisdiction in which your company is located. However, taking the federal Defend Trade Secrets Act enacted in 2016 as an example, one universal requirement for trade secret protection is that the owner took reasonable measures to maintain the secrecy of information that derives value from not being generally known. What may have constituted reasonable measures 10 years ago may not be found to be reasonable measures today, with the release of ever-more sophisticated IT and IP security technologies, necessitated by the relentless evolution of the sophistication of bad actors that try to steal IP. What is reasonable will of course, vary based on both the type and size of the company you are in charge of protecting, but all companies, regardless of size and industry, can benefit from developing a well-planned IP protection strategy (Fig. 1).

The bad actors that attempt to steal IP can be located anywhere in the world, and the barrage of cyberattacks on companies never ceases. The U.S. Cybersecurity & Infrastructure Security Agency (CISA) is a government agency that connects industry and government stakeholders to provide resources, analyses, and tools "to help them build their own cyber, communications, and physical security and resilience" (Cybersecurity and Infrastructure Security Agency 2023). In 2022, CISA issued a Cybersecurity Advisory regarding "Ransomware Awareness for Holidays and Weekends," warning that both the FBI and CISA "have observed an increase in highly impactful ransomware attacks occurring on holidays and weekends, when offices are normally closed, in the United States, as recently as the Fourth of July holiday in 2021" (Cybersecurity and Infrastructure Security Agency 2022). The

 
W. Harding et al., *Securing a Connected Future*,
https://doi.org/10.1007/978-3-032-07309-9_2

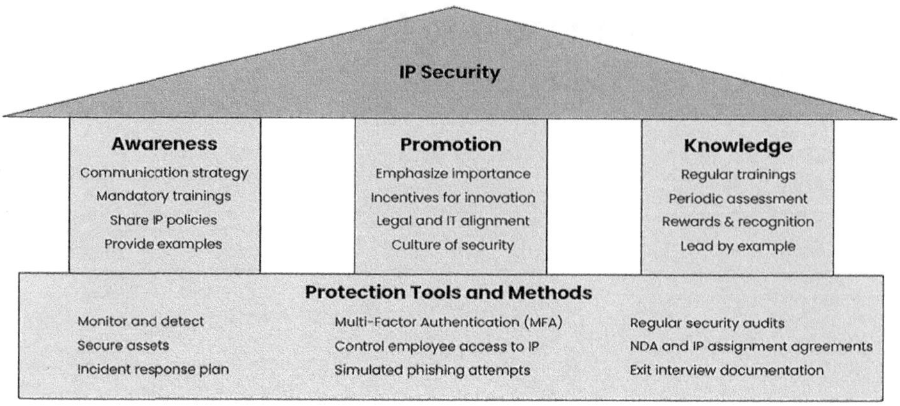


**Fig. 1** IP Law & cybersecurity

reasons for these spikes in attacks are that cyber criminals consider non-workdays as the perfect timing to single out desirable targets. Cybersecurity and threat detection companies have confirmed this trend. For example, Darktrace, a cyber security AI company, reported that its security researchers "discovered a 30% increase in the average number of attempted ransomware attacks globally over the holiday season in every consecutive year from 2018 to 2020 compared to the monthly average" (Darktrace 2021, para. 1). This makes sense, malicious actors are counting on the fact that many companies only have skeleton crews in place and so may be less vigilant or simply have fewer IT security professionals to prevent attacks.

But threats to IP protection can come from ***inside*** a company as well. A quintessential example, seen in myriad lawsuits every year, is a former employee of Company A who takes confidential and proprietary information with them to Company B, and then works with Company B to create a competitive product that undercuts Company A's profits, not to mention its years or decades of investments. Given both external and internal threats to IP protection, Chief Information Security Officers (CISOs) must exercise constant vigilance, and to do that, they need the buy-in and cooperation of every employee at their company. Even one "weak link" in a comprehensive web of cybersecurity and IP protection tools can allow IP to slip out of the company or bad actors to slip into the company's network and then move laterally, so it is vitally important that employees at every level, whether involved in the creation of IP or not, is both aware of the company's IP protection measures, and is willing and able to comply with them. Additionally, as a CISO, it is crucial to address board-level concerns that have the potential to impact the organization significantly. Several key questions that board members may have include:

- Could our trade secret IP be exposed using AI-enabled tools that might derive the details of our trade secrets? Examples of data that might be used are employee skills data, social communications, product BOMs, reverse-engineered products, supplier lists, etc., all of which might form a picture of what trade secret processes and designs are in place that enable the production of products.

- Could a competitor's AI tools be used to generate IP that is similar to mine, such that the competitor's AI-generated IP does not infringe on my IP, but does undermine my IP, potentially costing me market share while reducing my revenue? That question speaks to the need to plug holes in any developed IP so that competitors are not able to steal ideas.
- Are our internal systems secure enough that an AI-enabled hack could not gain access to our sensitive IP, such as patents being developed or trade secrets that are currently implemented? Of course, that question should be considered when developing any cybersecurity solution, but it does speak to the need to add more security around areas where IP is developed (e.g., innovation/R&D rooms) and systems where IP documentation is stored.

The following section provides an organizational change management framework to help organizations understand how to change and adapt their tools and practices to answer and address the questions above.

## 2 Adopting an OCM Strategy to Foster a Culture of Protection

As organizations consider adopting new policies to protect their IP and trade secrets, it is essential to take a proactive and strategic approach. Far too often, organizations implement new measures without an effective awareness approach, communication strategy, or training program, leading to failures. Cyber hacks target companies daily, and bad actors are becoming smarter and more strategic. For example, according to the Verizon 2023 Data Breach Investigations Report, a prominent healthcare provider suffered a significant data breach due to inadequate cybersecurity protocols. The breach exposed sensitive patient information and proprietary research data, resulting in substantial financial losses and damage to the organization's reputation. This incident could have been mitigated if the organization had employed a robust Organizational Change Management (OCM) strategy. By ensuring that all employees were aware of the new cybersecurity measures, desired to support the changes, were knowledgeable about their roles in maintaining security, capable of implementing the protocols, and continuously reinforced the importance of these measures, the organization could have significantly reduced the risk and impact of the breach.

A structured and centralized approach is the first step in adopting new policies to protect intellectual property. Many initiatives with good intent fail due to a lack of ownership, resulting in finger-pointing and chaotic situations. A proven method to ensure organizational readiness for change is the Awareness, Desire, Knowledge, Ability, & Reinforcement (ADKAR) methodology. This method, used by global organizations, ensures employees are prepared to understand and implement new cybersecurity measures effectively.

## 2.1 Who Should Be Involved?

Executive stakeholders and sponsors should be involved from inception through execution and ongoing sustainability. Leaders must drive the communication and underscore the importance of the change. Without leadership's support, employees are often resistant to change. Additionally, a central point of contact or function, such as a Project Management Office (PMO) or Center of Excellence (CoE), is essential for accountability and success. This function drives program updates and keeps the leadership team informed, ensuring that the change initiative remains aligned with organizational goals and that any obstacles are promptly addressed.

## 2.2 Why Change?

There are foundational elements to establish when adopting new changes. Protecting intellectual property and preparing for cyber threats requires a mindset shift. Employees must agree to the changes, which can disrupt long-standing routines. Establishing a 'Case for Change' communicates the value behind the change and aligns it with executive stakeholders. Examples of reasons for change include:

- Identifying Specific Vulnerabilities: Understanding the current gaps in IP protection measures.
- Assessing Past Impacts: Evaluating how previous breaches have affected the organization financially and reputationally.
- Recognizing External Threats: Identifying the most relevant cyber threats specific to the industry.
- Highlighting Urgency: Explaining why immediate action is necessary.
- Anticipating Benefits: Outlining the advantages of enhanced security and regulatory compliance.
- Aligning With Strategic Goals: Ensuring that the changes support long-term organizational objectives.

As an organization investigates establishing the reasons behind a change, it helps to address challenges, such as employees not understanding the value of the change. Employees are resistant to change, and unless they have clear direction and understand the benefits, it will be difficult to persuade them to alter their current practices. For safeguarding cybersecurity, it is beneficial to identify the reasons for change.

## 2.3 Operational & Organizational Changes

Implementing new cybersecurity protocols often necessitates changes in operations and the organizational environment. This might include updating software systems, revising workflows, and enhancing physical security measures. It is crucial to outline these changes meticulously to minimize disruptions. For instance, introducing multi-factor authentication (MFA) may require alterations in how employees access systems daily, or AI-based anti-phishing email systems may require training and education. By clearly outlining these operational changes, organizations can ensure a smooth transition and maintain productivity during the implementation phase.

## 2.4 Change Implementation

Having a change implementation or OCM Action Plan is vital for the successful adoption of new cybersecurity measures. This plan should detail the steps required to transition from the current state to the desired future state. Key components of an effective implementation plan include:

- <u>Timeline</u>: Establish clear milestones and deadlines for each phase of the change. Where appropriate, map the Work Breakdown Structure (WBS) to corresponding stages in your OCM methodology (e.g., ADKAR or Prosci) to ensure alignment between project execution and change adoption.
- <u>Resources</u>: Allocating necessary resources, including budget, personnel, and technology.
- <u>Accountability/Roles and Responsibilities</u>: Defining roles and responsibilities to ensure accountability of the program.
- <u>Training Programs</u>: Developing and scheduling training sessions to equip employees with the necessary skills and knowledge.
- <u>Communication Strategy</u>: Outlining how updates and progress will be communicated to stakeholders.

By following a structured implementation plan, organizations can address each aspect of the change, ensuring that no critical elements are overlooked.

## 2.5 Success Metrics/Key Performance Indicators (KPIs)

To monitor the effectiveness of the change initiative, it is essential to establish success metrics and Key Performance Indicators (KPIs). These metrics provide tangible evidence of progress and highlight areas that may require additional attention. Examples of relevant KPIs include:

- Employee Adoption Rates: Measuring the percentage of employees who have completed training and are actively using the new cybersecurity protocols.
- Incident Response Time: Tracking the time taken to respond to and mitigate cybersecurity incidents.
- Compliance Rates: Assessing adherence to new policies and procedures.
- Security Breaches: Monitoring the number and severity of security breaches post-implementation of cybersecurity measures.
- Employee Feedback: Gather qualitative data through surveys and feedback mechanisms to gauge employee sentiment and identify areas for improvement.

Reviewing these metrics allows executive sponsors and program leaders to make informed decisions, adjust strategies as needed, and ensure that the change initiative remains on track.

## 3 From Theory to Practice: Enabling Change Through ADKAR

While various OCM methodologies exist, ADKAR is one of the most widely recognized frameworks for guiding individuals and organizations through change. Some organizations may adopt Kotter's 8-Step Change Model, while consulting firms often develop proprietary approaches based on their experience. However, when implementing technology-driven change, especially in cybersecurity, organizations must tailor their approach to align with the specific risks, systems, and adoption challenges unique to their environment. In this section, we will explore how the ADKAR methodology can be effectively applied to a cybersecurity initiative, ensuring that new security protocols are not only implemented but fully adopted and sustained across the organization (Fig. 2).

### *3.1 Awareness Purpose*

Building upon the 'Why' exercise, the awareness phase ensures that all employees understand the critical need for enhanced cybersecurity measures. Often, organizations overlook the importance of clearly communicating the reasons behind such changes. For example, sharing the details of a data breach and its impact helps teams understand the severe implications a breach can have on the organization. This knowledge sharing not only highlights the necessity for change but also motivates teams to actively support and implement the new cybersecurity measures. By openly sharing these insights, organizations can ensure that everyone is aligned and committed to safeguarding their organization's IP. Ensuring that there is a feedback loop, such as Q&A sessions, surveys, and open forums, helps address areas where employees may not understand how to implement the strategy.

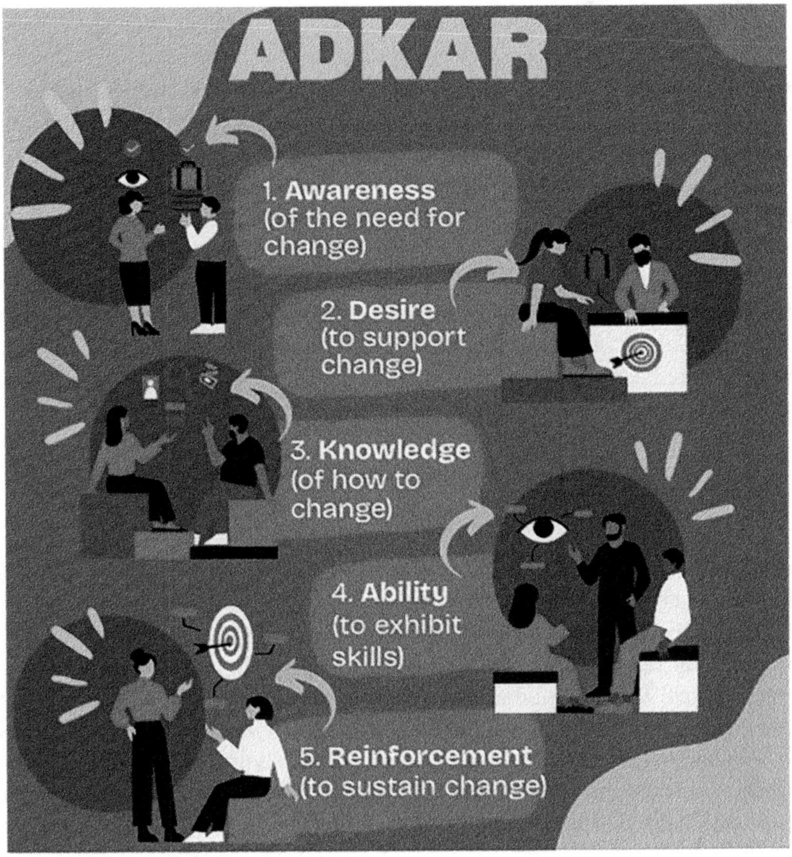


**Fig. 2** Organization change through ADKAR

## 3.2 *Desire Purpose*

Foster a desire to participate and support the change. This involves creating a motivation and advocacy plan for employees to embrace the new cybersecurity measures. Companies should engage employees by involving executive leaders and subject matter experts (SMEs) in the communication plan, presenting data that underscores the importance of the change. These are team members who may become change champions to support the change. Additionally, involving employees in planning and decision-making processes (such as by forming focus groups or task forces that include representatives from different departments) can help tailor the change to meet diverse needs. Lastly, recognizing and rewarding early adopters and supporters, such as by awarding certificates or public acknowledgment in company meetings, can incentivize employees to embrace the change.

## 3.3 *Knowledge Purpose*

Provide the knowledge and training needed to implement the change so that employees understand not only what the new protocols are but also how to execute them effectively. This will require developing comprehensive training programs covering new policies and procedures will ensure widespread understanding, as will offering access to resources, such as guides, FAQs, and help desks. Utilizing platforms like SharePoint or an equivalent tool can centralize these resources, making them easily accessible to all employees. Lastly, it is important to ensure ongoing support through mentoring and coaching. Establishing a Community of Practice (CoP) or Center of Excellence (CoE) can provide a support network where employees can seek assistance and share best practices.

## 3.4 *Ability Purpose*

Equip employees with the skills and capabilities to apply the change. Practical application is crucial for the successful implementation of new cybersecurity measures. Using simulations, role-playing exercises, and sandbox environments where employees can safely apply new protocols will provide opportunities for hands-on practice and real-world application. Even after new policies have been implemented, it is important to continue to conduct assessments to identify skill gaps and address them. Practical tests or quizzes can help evaluate employees' understanding and readiness to implement the changes effectively, and regular check-ins and refresher courses can help maintain high levels of proficiency and address any ongoing challenges.

## 3.5 *Reinforcement Purpose*

Ensure the change is sustained over time. Reinforcement strategies help maintain momentum and embed the new cybersecurity measures into the organizational culture. Sustained change will require establishing metrics to monitor progress and adherence. Regularly reviewing KPIs, conducting audits, and collecting and acting on feedback from employees can help ensure that the changes are being maintained effectively. Lastly, it is important to celebrate milestones and successes to reinforce the change. Recognizing achievements, whether through awards, public acknowledgments, or team celebrations, can help sustain enthusiasm and commitment.

# 4 Change Modules & Tools

To facilitate the successful adoption of new cybersecurity measures, various change tools can be employed. These tools support different stages of the ADKAR methodology and help streamline the change process.

## *4.1 Change Catalyst Sizzle Reel*

Create a short, engaging video that highlights the key benefits and goals of the new IP protection measures. This tool can be used to build excitement and buy-in from employees and stakeholders. Share it during kickoff meetings, company-wide emails, and on the company intranet to maximize reach and impact.

## *4.2 Change Impact Assessment*

Identify how changes will impact different parts of the organization. This assessment helps leaders understand and plan for necessary adjustments, ensuring that all departments are prepared for the transition.

## *4.3 Communication Plan*

Ensure consistent and clear messaging throughout the organization. A well-structured communication plan maintains transparency and keeps everyone informed about progress, upcoming milestones, and any changes to the implementation strategy.

## *4.4 Training Programs*

Equip employees with the knowledge and skills needed for the new processes. Comprehensive training programs enhance employee competence and confidence in the new measures, ensuring they can implement them effectively.

## 4.5 *Stakeholder Engagement*

Gain buy-in and support from key stakeholders. Engaging stakeholders early and addressing their concerns ensures alignment and fosters a collaborative environment for change.

## 4.6 *Feedback Mechanisms*

Collect input and monitor employee sentiments through surveys, focus groups, and suggestion boxes. This feedback allows organizations to adjust strategies based on employee insights and improve adoption rates.

## 4.7 *Reinforcement Strategies*

Ensure long-term adherence to new processes. Reinforcement strategies, such as regular updates and recognition programs, help maintain momentum and encourage continuous improvement.

## 4.8 *Change Champion Network*

Leverage influential employees to drive change. Change champions facilitate peer support and encourage positive attitudes toward the new cybersecurity measures, helping to foster a culture of security.

## 4.9 *Visual Aids and Infographics*

Simplify complex information and make it accessible. Visual aids enhance understanding and retention of information, making it easier for employees to grasp the new protocols.

### 4.10 Regular Updates, Newsletters, and Spotlighting Success Stories

Keep the organization informed about progress and upcoming milestones. Regular updates maintain engagement and transparency throughout the change process, ensuring that employees remain informed and motivated.

Whether your organization decides to protect its intellectual property through patents, trade secrets, or a combination of both, putting in place a robust IP protection plan is a critical step for all board members and CISOs. The significant financial and human capital investments in the creation of IP are clear, and efforts to earn a return on that investment and prevent employees and third parties alike from running roughshod over your company's IP rights are a never-ending and constantly evolving challenge, but a necessary one, and one that is worth it. The suggestions above provide a high-level framework for CISOs at all types of organizations to monitor IP from the moment of conception through all steps of the protection process, and beyond. By leveraging the ADKAR methodology within an effective Organizational Change Management strategy, medical organizations can ensure that their cybersecurity protocols are not only adopted but also sustained, thereby safeguarding their most valuable assets against ever-evolving cyber threats.

## 5 Conclusion

Employees are not just part of the IP protection strategy; they are responsible for executing it. Raising awareness and building desire across the organization is essential to creating a culture of ownership around IP protection. With risks coming from both external threats and internal vulnerabilities, effective protection requires a continuous blend of people-driven vigilance and evolving technology. Success comes from having a robust OCM approach and partnering with the right vendors who understand both the threat landscape and the importance of adoption. The next chapter will explore these elements in greater detail.

# Best Practices for Protecting IP & Ensuring the Security of Medical Devices

## 1 Translating IP Protection Policy into Everyday Practice

Before you can effectively implement an IP protection policy, every employee, from executives to entry-level staff, needs to be aware that such a policy exists. A crucial part of ensuring that the staff understands the importance of IP protection is by developing a communication strategy that is tailored to targeted user and shareholder groups. Tailoring the messaging to the audience will also benefit knowledge retention since there will be varying levels of technical knowledge among these groups. For example, executives may require high-level information, such as explaining the need to protect IP and how IP protection helps the company maintain a competitive advantage. Executives may also be interested in the revenue generation and brand protection aspects of strong IP protection measures. A company's IP is its crown jewels, often requiring years or decades of significant investment of human and financial capital. Meanwhile, non-executive staff, who may be more involved in day-to-day operations and handling data and IP assets, may need more hands-on training associated with the technical aspects of IP protection, as well as the higher-level thematic reasons why it is important.

Mandatory training, both proactive and remedial, is important for educating employees on any changes to the IP protection policies, as well as through repetitive training on current policies. Ultimately, enhancing awareness, regardless of department or role, is crucial for the success of a holistic IP security training program. Promoting an organization's policies and highlighting the potential risks associated with IP security breaches will help employees understand the importance of such measures. These training programs also establish clear expectations that IP security is a top priority for the company.

A robust training program should also include real-world examples that help staff understand the importance of protecting an organization's IP and following the provided IP security measures. For example, every year, multiple jury verdicts



W. Harding et al., *Securing a Connected Future*,
https://doi.org/10.1007/978-3-032-07309-9_3

award hundreds of millions, or even billions, of dollars in damages to companies that were lucky enough to detect and prove patent infringement and/or trade secret theft. As just one recent example, a former employee of a cloud computing and enterprise software company took a copy of that company's proprietary and confidential software to a competitor and then conspired with the competitor to create its product to compete with the plaintiff's product (Proskauer 2023).

Emphasizing IP protection policies may not be high on the average employee's list of priorities, but employee education on the topic of IP protection helps employees recognize the potentially catastrophic implications of IP security breaches. Thus, education reinforces the importance of implementing strong security measures to safeguard valuable information.

## 2 Stimulate Employees to Actively Participate in IP Protection

Emphasizing the company's substantial investment in intellectual property, both in employee-hours and financial capital, will stimulate an organization's understanding of the critical need for safeguarding IP as well as promoting a sense of responsibility among employees. Ongoing communication of IP security during employee onboarding, performance evaluations, internal communications, and exit procedures will ensure a culture of security. By highlighting the significant investments made by the company in intellectual property and its importance to the organization's success, employees can better appreciate the value and significance of protecting IP (Fig. 1).

One way to gain employee buy-in for IP protection policies is to foster pride of ownership of technology and inventions. Whether employees are entry-level inventors or have been developing technology for years, it is essential to make employees feel appreciated for their work. Further, it is important that employees understand how important their innovations and contributions are to the company's reputation as well as to the company's status amongst competitors, which helps employees feel like they are all working towards a common goal.

Another way to generate employee enthusiasm for IP protection is to create incentives for innovation and the associated protection of that innovation. Of course, technical employees like software and hardware developers are hired with the expectation that they will contribute to a company's technology portfolio, but providing an extra layer of recognition can help motivate them even more. For example, many companies implement patent incentive programs, through which employees can receive nominal monetary awards or other recognition for the various stages of their contribution to a patent, such as when a patent is filed, and again when a patent is granted. Providing tokens of recognition, such as plaques and certificates, or hosting a yearly dinner event to honor major contributors to the

**Fig. 1** Employees protecting IP (CollectiveCampus.io)

company's IP, can help ensure employees feel their IP contributions are valued and make them want to protect them even more.

William Harding, Ph.D., a prolific innovator at medical device company Medtronic for over two-and-a-half decades, provides an engineering employee's perspective on the human importance of IP creation and protection: "Working for a medical device manufacturing company should never be viewed as just a job. While we all want to make a living, provide for our families, and save for our future, the work that enables the creation of life-saving technology is more about making a difference and feeling appreciated for the work". Employee knowledge that the work they perform enables employees and companies to create devices that improve a patient's quality of life, as well as provide the tools that ease the stress and burden that healthcare professionals might experience as they seek to restore the lives of their patients, motivates the employee to continue creating needed IP. And part of making a difference to patients includes creating novel technological solutions that enable a company to lead the world in the design, development, manufacturing, deployment, and sustaining of advanced medical technology.

The creation of novel medical technology does not happen overnight; instead, it takes years, if not decades, of ideation, prototyping, testing, and iteration to arrive at a solution that meets the needs of customers. That creation process starts with establishing empathy with a patient's needs (or healthcare professional's needs), hearing their story, and putting a team of talented innovators in the same room. But merely throwing a bunch of smart and talented people into a room and hoping for success does not necessarily lead to innovation. Companies need to create environments that support and stimulate their innovators, and to train those individuals in how to work together and how to use ideation methods and tools like Design Thinking in a fast, iterative process that helps them brainstorm, capture, and demonstrate their ideas. That process requires a company to invest in the training of coaches and instructors who can guide the innovators through the process of generating new technological solutions.

Contrary to what certain executives may assume, a company's technology teams and engineers are ***not*** the only people who should be involved in the process of ideation. Instead, the most innovative concepts are often created through the inclusion of diverse perspectives from contributors in different departments and from teams composed of individuals who bring different views and experiences. Facilitating work in cross-functional teams is not always cheap, and companies invest significant employee time and financial capital throughout the process.

With time and money in mind, there is more to developing new IP than just the cost of the tools and environments, but also the time and even the incentives that might be offered to employees to motivate them to be innovative. Many companies make the mistake of placing a piece of paper in front of someone and telling them to create a solution that addresses some specific medical need. But those practices only result in a list of ideas, not prototypes or diverse perspectives, and they rarely address the actual needs of customers. That said, a list of ideas is unlikely to be ready to be filed as a potential patent, which ultimately leads to failures, rejections, and a waste of time and money. Thus, recognizing that creating and protecting IP requires significant initial investment is important, as the investment in robust ideation and IP protection procedures always has an ROI that makes it worth it.

Employees are not there to perform rote ideation tasks or to create solutions without incentives; expecting as much is likely to lead to a continuous need to backfill the positions that have been vacated by employees who left due to feeling unfulfilled and unappreciated. Instead, approaching IP creation and protection with the mindset that new concepts require a nurturing and rewarding environment will allow companies to witness the pride that employees experience in knowing that their company appreciates their efforts and recognizes that they have truly made a difference in the lives of their customers. In sum, the best way to encourage a culture of security within your organization is by promoting a sense of responsibility among employees. That point emphasizes the importance of IP protection by tying accountability and ownership to the company's overall goal of improving the lives of its customers. It is a rare honor to be listed as an inventor on a patent, and innovative employees should be rewarded and recognized for their contributions and demonstrated commitment to IP security.

## 3 Educate Employees to Protect IP

Once a company has established best practices for motivating its employees to ***create*** IP, it needs a plan to train its employees to ***protect*** IP during its development phase. After an organization's staff has received IP protection training, it is crucial to provide them with a support structure for continuous learning and knowledge retention. Ongoing training modules should be tailored to the specific needs of various departments within an organization. For instance, an engineering group may heavily focus on the various tools and methods to effectively safeguard the organization's intellectual property. And by offering department-specific IP protection training modules, employees can dive deeper into the specific challenges and requirements of their roles. The training also helps the employees stay up-to-date with the latest best practices and techniques. That said, security measures must always evolve, because the bad guys (including those acting on behalf of nation states) are always evolving. Accordingly, it is important to follow a robust IP security framework to ensure that your company is using the most effective and up-to-date security measures.

Following training, the creation of IP (e.g., patents and trade secrets) combines several tools, practices, coaching, incentives, and an awareness of security measures that result in novel and foundational solutions. Thus, incorrectly implemented and managed IP generation/capture processes can result in IP being compromised, which increases the risks of lost market share within industries that consider IP to be their bread and butter. For example, the following list represents the principal elements that enable a company to effectively create and secure IP.

- Design Thinking Coaches: Individuals trained in methods and the use of tools that enable a company to quickly ideate on new concepts.
- Internal IP Training Courses (e.g., IP 101, TRIZ, Preceptorship Programs): Internal and external programs that are used to train innovators in the creation of IP and the research into existing IP, which ensures that the IP is novel/seminal. It also involves providing innovators with access to their customers.
- Internal Ideation Exercises: Secure internal conferences and programs that are offered to stimulate new ideas and develop empathy with the stories/needs of their customers. It is important for in-house counsel to be involved in managing the outcomes of ideation exercises as well, to document the conception and reduction to practice of inventions that may eventually be filed as patents. The legal team can also play a role in ensuring that only employees who need to be present at the ideation sessions participate, and that third parties and employees without a need to know are not needlessly exposed to ideas stemming from the discussion (such as pictures taken of whiteboards, or through unnecessary discussions).
- IP Capture Documents: Physical documents that are used in the capture of ideas, before they are entered into a digital IP collection system, should be kept secure (e.g., in locked file cabinets), not lying around on an employee's desk.

- Low-tech Prototyping Material: Inexpensive material that is used by innovators to quickly prototype their ideas, to share ideas, to solicit feedback, and to test those ideas quickly and without much expense. At the same time, employees in charge of prototyping should be made aware that they are also responsible for maintaining the confidentiality of those prototypes by keeping them in secure areas (like a locked file cabinet) away from casual passers-by (including other employees and third parties alike).
- IP Capture Software/Websites: Secure systems that enable an innovator to digitally collect their written ideas and prototype diagrams/photos into a format that can be reviewed and filed. The system should enable the innovator to track their IP, to designate where the IP is assigned (i.e., which organization within the company), and to receive feedback associated with their IP.
- IP Review Boards: Internal teams composed of lead innovators, leadership, and legal representatives. The innovator will present their ideas to the appropriate board and receive feedback as well as encouragement that will help them improve and move their ideas along the path to becoming a patent or trade secret.
- Legal Advisors: After working with review boards, which should also have legal representatives, the innovator will work with the internal legal department to refine the approved idea until the full extent of the idea is captured. The work with the legal team will continue as the IP is either publicly disclosed in a patent application or until the decision is made to maintain that IP as a trade secret and ***not*** to publicly file it in a patent.
- Monitoring Tools (i.e., Related to the Movement of IP Across Software Platforms): These tools are not often known to the innovator, but are put into place as an additional security measure to ensure that IP is not being shared outside the confines of a managed IP protection system. A more in-depth discussion of monitoring tools is included in Section D.

Many innovators use an assortment of technologies in the creation of IP, but to ensure organizations are thoughtful in selecting the right technology for IP security, the following checklist provides a structured approach to making targeted, strategic investments.

## 4 IP Security Needs Analysis

- Define the specific IP and cybersecurity risks or other technologies that will help resolve them.
- Align technology goals with cybersecurity and IP business outcomes using measurable KPIs.
- Engage experts (e.g., solutions architects, consultancies) to assess and determine the best-fit technology.

## 5 Technology Selection

- Assess vendor financial stability when selecting a startup to ensure its sustainability beyond your business.
- Verify technical certifications required for industry compliance (e.g., ISO certifications).
- Ensure the technology integrates seamlessly with your existing security and IT infrastructure.

## 6 Implementation and Adoption

- Partner with a professional services organization with proven success, and check for testimonials and case studies.
- Avoid "cool but useless", define real business use cases, and keep OCM and end-user needs at the forefront.
- Involve the business early and do not let IT drive transformation in isolation. Engage key SMEs with influence in the business throughout scoping and implementation.
- Integrate technology with existing security tools and infrastructure to avoid silos.

## 7 Iterative Impact Measurement

- Use the KPIs established during scoping as your North Star, while continuously tracking and evaluating adoption.
- Showcase successes and highlight positive outcomes in QBRs as well as across the organization, to build momentum for future rollouts.
- Measure reductions in cybersecurity/IP risk and improvements in response times.
- Implement governance to refine and improve AI models continuously.

## 8 Vendor Selection for Long-Term Success

Within the scope of creating and securing new IP, it is often the case that outside vendors are stakeholders in all phases of IP creation. And as organizations evaluate new technologies, it is essential to assess a vendor's fit, domain expertise, and long-term viability. Partnering with a smaller or emerging vendor can offer agility and flexibility, but also introduces risk, especially if the company lacks financial stability or operational maturity. As noted in the technology selection checklist, it is crucial to confirm that a vendor has the needed funding and structure to support the technology beyond initial deployment. The worst-case scenario is investing in a solution that becomes unsupported or deprecated mid-implementation.

Here are five key areas to assess when selecting a vendor for a potential partnership:

- Historical Success and Industry Alignment: Reviewing a vendor's track record provides insight into their ability to deliver. Look for tangible success stories, ideally from organizations in your industry. Case studies, press coverage, or analyst validation can help confirm whether the vendor understands the unique compliance, security, or IP protection challenges your organization faces.
- Team Expertise and Domain Knowledge: Vendors with deep subject matter expertise (SMEs) in your industry bring more than technology, where they also bring context. While some vendors rely on third-party GSIs or boutique firms to implement their technology, those with internal industry-focused teams tend to deliver greater continuity and insight. These teams know how to scope, implement, and sustain the change in environments like yours.
- Customer Success as a Strategic Function: A motto I have often used is: 'Our customers' success is our success.' In a world of rapid technology shifts, adoption is the new metric of value. Organizations should look for vendors with a formal customer success function, which is not just focused on retention, but on change enablement, adoption, and value realization. That includes ensuring CSMs (Customer Success Managers) are skilled in Organizational Change Management (OCM) principles and can help customers embed new tools into their workflows. The new mindset is not just about providing support, where it is also about proactive coaching and continuous improvement.
- Governance and Leadership Engagement: Technology programs do not succeed by chance. They succeed through intentional governance. Executive and Quarterly Business Reviews (EBRs and QBRs) are not 'nice to haves', they are essential 'must haves'. These reviews should include leadership from your team, the vendor, and any implementation partners to ensure strategic alignment, accountability, and course correction as needed. Robust governance creates space to assess whether the technology is meeting expectations and when to explore additional value-added capabilities.
- Value Realization and North Star Alignment**:** All of this ultimately leads to value realization. That starts on Day 1. Before the first line of code is written or the first configuration is made, organizations should define the North Star goal, the guiding business objective that the technology is expected to support. When this goal is clear, it becomes easier to align change management efforts, measure success, and adjust as needed. Without this alignment, even the best technology can become shelfware. You can have the best technology, but without an OCM-centric approach, the most innovative technology implementations sometimes fail.

By embedding an OCM-centric Customer Success model into your vendor selection and implementation strategy, your organization ensures that technology adoption is driven by user enablement, measurable outcomes, and long-term value, not just go-live success.

# 9 Developing a Robust Portfolio of IP Protection—Tools and Methods

A single patent or trade secret is not enough to drive an organization's success. And for that reason, most companies invest money, time, and resources into developing and strengthening a philosophy of continuous IP development. Moreover, the need to develop a portfolio of IP also requires the development of an IP protection portfolio. And with both portfolios in mind, the following tools and methods are offered for consideration when looking to expose security risks.

## *9.1 Electronic Measures*

**Monitor Download Activity** In many (but not all) cases, an employee's mass downloading of documents could be a red flag that that employee is accessing those documents for an improper purpose, such as to save them into an external drive (e.g., USB) and upload them somewhere else. Of course, certain employees may need to move large numbers of files as part of their job duties, in which case their mass download activity may not indicate anything nefarious. Thus, while it can be informative to understand how many documents particular employees are accessing and/or downloading, an even better data point is to understand if certain access/ download activity is *abnormal* for that employee. There are numerous electronic tools available that can alert IT security professionals when access/download activity exceeds a certain level, and/or when abnormally high access/download activity is detected.

**Audit Passwords and Permissions** In addition to requiring robust passwords (and the regular changing of those passwords), organizations should consistently reassess permissions provided to employees to confirm that they are needed. Certain employees simply do not need access to certain types of information (such a source code, technical specifications, etc.), and it is best to cut down on the risk that someone without a need to access those documents or source code can access them for an improper purpose. Moreover, two-factor authentication should be the norm for computer access and access to different internal databases, thus helping prevent further lateral movement if a bad actor can circumvent a first level of defense, such as by gaining an employee's password (IEEE/UL, 2024, Standard no. 2933-2024).

**Install Threat Detection Software** There are dozens of state-of-the-art threat detection tools that companies can buy or license. The best threat-detection tools are frequently updated, even in real-time, to disseminate information about current known threats, whether coming from specific IP addresses, websites, or even countries. These tools can allow companies to prevent the threats from ever reaching their networks, or if they do, to detect the malicious activity and compromised machines quickly enough to prevent lateral movement through the network. The importance of threat detection software is even more pronounced given the number of employees working fully or partially remotely on home networks that are likely not subject to the same regulations and security measures as on-premises networks.

**Mandate Recurring IP Policy Training** One training on IP protection when an employee joins a company is simply not enough. Employees need to be refreshed on existing policies and to receive updated information on new or different procedures. IP protection training should be mandatory, with consequences for failing to complete it correctly and on time. And as described further in the "In Person Measures" section, reminding departing employees of their ongoing IP protection obligations is also an important step.

**Deploy Mock Phishing Attacks on Employees** While it may seem counterintuitive to phish your employees, these controlled tests can provide important insight into the efficacy of your company's IP training and provide remedial education to employees who need it. The IT department can randomly send emails that either spoof an executive's email address and ask an employee for something they should not be providing or include a suspicious link that should not be clicked. Employees who fall for this simulated phishing attack can be directed to complete remedial training to help them understand how to avoid such pitfalls in the future, and to understand the consequences that could have befallen them and the company had the attack been real.

**Backup Your Data Offline** CISA also provides numerous recommendations and best practices for avoiding or overcoming ransomware attacks, and in addition to the more obvious recommendations of not clicking on suspicious links, regularly updating your operating system and software, scanning for vulnerabilities, and using strong passwords, CISA also recommends making an offline backup of your company's data (Cybersecurity and Infrastructure Security Agency 2022). By conducting regular backups of data and storing those backups *offline*, your company will make it harder for cyber criminals to hold your data hostage or delete online backups. By using this tactic, your company is less likely to have to make the unenviable decision between losing potentially all your company's IP (like source code, technical specifications, and other technical documents developed through significant investments of time and money) or paying an exorbitant ransom (which does not even guarantee the return of your data).

## 9.2 In-Person Measures

**Cut Off System Access When No Longer Needed** When an employee gives notice that they intend to leave the company, or the company decides to lay an employee off, deciding when and how to cut off that employee's access to internal repositories is a balancing act. Recently, companies have been criticized for suddenly cutting off employees' access to Slack and other internal systems without first notifying them that they had been laid off. But on the other hand, allowing employees to continue to have access to technical documents and source code repositories in the weeks or days leading to their departure can provide a window of opportunity for employees to download and take documents they should not. If immediately cutting off access is not a viable option (e.g., because employees need to transfer their current projects to others and upload existing technical documentation before their computers are wiped), then best practices would include ensuring that the departing employee's access and download activities are, at a minimum, monitored and that the IT department would get an alert about unusually high or unusual activity.

**Perform Security Audits** IT security and IP protection measures must always be evolving, which requires periodic audits of existing procedures to identify any points of weakness and to update tools and policies accordingly. For larger companies, it may make sense to hire an external company to perform an audit and provide the IT department with a detailed report with scores for different measures of IT security, along with suggestions for bolstering the effectiveness of the IT and IP security program.

**Require Non-disclosure Agreements (NDAs) and Confirm Return of IP Upon Departure** NDAs can be an important tool not only when an employee joins the company, but also when they leave the company, to remind them of their continuing obligations not to use or disclose the company's IP, even as they join a new company. After all, just because a former employee can remember a trade secret in their mind, without reference to a document, does not mean that they are free to use that trade secret at a new company. Thus, requiring the signing of a NDA upon an employee's exit from the company can help remind them of the specific technology they worked on at your company, and clarify that they are not entitled to retain confidential and proprietary information (whether in their mind or written down), documents, prototypes, etc., and should return all company-owned intellectual property to the company. Most departing employees *want* to do the right thing and want to comply with their obligations, and conducting an NDA refresher upon the employee's exit is one way to help them comply.

### *9.3 Legal Measures*

**Patents** One primary way for companies to protect their innovations is to file patents on their employees' inventions. The path from filing a patent to having it granted can be a years-long process, often with multiple back-and-forth communications between a company's patent lawyers and the U.S. Patent and Trademark Office (USPTO). One recent study has found that most patent examiners issue three Office Actions (e.g., rejections) before granting a patent (Tu 2018). But this is not to say that the patent process is not worth it. A U.S. patent is typically valid for 20 years from the filing date, during which time the patent owner can exclude all other (U.S.) companies from making, using, or selling the patented product or process, and can sue another company for lost profits (among other remedies) if that other company infringes.

**Trade Secrets** Trade secrets are another major tool for a company to protect its IP. In some ways, a trade secret is the opposite of a patent. A patent owner decides to make its innovative product or process publicly known through the filing of a patent application, while the owner of a trade secret does not need to register anything at all and can instead choose to simply maintain the secrecy of information that derives economic value from. Thus, in theory, contrary to a patent that expires in 20 years, a trade secret could be maintained in perpetuity. For example, the formula for Coca-Cola was never patented and has instead been maintained as a company trade secret for more than 130 years, and to this day, no other company has been able to replicate it (Cola 2020). Similarly, KFC's blend of 11 herbs and spices has been maintained as a trade secret since it was created in 1940; to maintain it as a secret, two different companies assemble the mixture, such that no single third-party is privy to the entire recipe (Myers 2018). But as a cautionary tale for how easy it is for trade secret protection to be destroyed forever in an instant, a scrapbook of a Colonel Sanders family member containing a recipe for a blend for 11 herbs and spices was shown to a reporter for The Chicago Tribune with a claim that it was the secret recipe, resulting in the company issuing a statement that it was the *not* the correct recipe (Stack 2016). This near-catastrophic competitive loss of a trade secret for KFC merely highlights how critically important it is for companies to prevent leaks of IP through the education of employees, reinforced with robust technical protections.

## 10 Organizational Special Considerations

Moving from evaluating security risks, a 2021 study was conducted by Boston-based cybersecurity firm Cybereason on over 1000 companies that had experienced a ransomware attack during a holiday or weekend. The study exposed that companies in the healthcare industry were one of two sectors ***least*** likely to have developed a contingency plan for a cyber-attack and nearly 75% of all survey company employees "admitted they were intoxicated while responding to a ransomware attack on the weekend or during a holiday" (Sganga 2021b, para. 9). According to cybersecurity firm Mandian, certain ransomware threats are specifically targeting the data of the healthcare industry (Shilko et al. 2021). Those cyberattacks on the data of healthcare organizations have led to adverse patient outcomes (Sganga 2021a), where it was also shown that the negative impacts of the attacks could last for years.

To help healthcare organizations identify what data needs to be protected, HIPAA regulations introduce an additional layer of complexity. HIPAA law protects patient health information, including any self-identifying information (e.g., patient age, email address, medical account numbers, device serial numbers, etc.). In order to meet HIPAA requirements, considerations need to be made for the protection of patient information, which may include organizational ideas (i.e., IP) regarding how patient data is protected/encrypted. During the process of developing a new medical device, it is often the case that the innovator only considers solving the immediate healthcare needs of the patient or healthcare professional. For example, if a patient needs a device that improves heart function, there is little to no initial consideration for ensuring the device is HIPAA compliant. But as the idea evolves into an actual device, and as more stakeholders become involved in the process, members from Engineering, IT, QA, and Regulatory groups will inject their input so that the device will comply with the FDA, TÜV SÜD, and HIPAA regulations (to name a few). It is at that time, and before the device is manufactured, that steps will be taken to ensure that data captured, generated, transformed, and transmitted by the device remains secure. Consequently, before a new device is even tested on a human, all members of the device development team have explored potential risks and have taken the greatest care to confirm that the device meets all government requirements.

It is normal to think that most medical solutions manifest as hardware; however, there are exceptions to that rule, where many solutions are purely software. For example, in developing a patented metadata encapsulated token authentication solution (i.e., influenced by blockchain concepts), the intent was to address the need to protect patient data, but not necessarily the method used to protect the data. Specifically, the desire was to design a solution that would assure HIPAA compliance, even though the data protection methodology would be shared with other organizations that seek to implement the same level of security around their protected content. And as reflected in Fig. 2, company-owned content, software, methods, and solutions can be shared openly without worrying about a loss of control over that content. That said, the now patented solution was designed to be more secure than would be expected by HIPAA, and because the solution was based on a blockchain model, steps were taken to ensure that even if transmitted data were intercepted, the data could not be decrypted. This security solution example represents a globally available model that includes emerging technology and methods of development that should be in the forefront of all healthcare related solutions.

**Fig. 2** Open source sharing with protection (Openverse.Org)

## 11 Conclusion

An organization's employees are some of the most important resources in the battle to reduce security risks and to protect highly prized IP. A well-trained employee, who is aware of security threats and motivated to defeat them, is the best tool in any cybersecurity toolbox. On the first day of any employee's training, they should be made aware of the important role that they play in protecting IP, the company's crown jewels. The tools and methods for protecting IP described in this chapter detail the steps and actions that organizations should follow when selecting cybersecurity technology, as well as vendor partnerships.

# Device Security & Standards

## 1 Medical Device Product Security

### *1.1 Future Forward Risks*

Looking into the future, medical devices will continue to become more integrated across networks and linked to other medical devices. This includes wearables, such as smart watches and blood pressure monitors, to implantables (e.g., pacemakers), and electronic health records. Additionally, AI will be used to predict and detect abnormalities to get treatment more quickly. While this functionality will be a great benefit to the patient and change the way of medical treatment, it will also pose an increased threat to product and software security. Some areas of concern to consider include ransomware attacks, device manufacturing attacks, insufficient device disposal, and AI/ML-enhanced attacks, as well as software vulnerabilities.

Ransomware attacks have occurred across various sectors, such as Stillwater Mining Company, Kawasaki Motors Europe, Disney, and even the city of Columbus, OH. That said, in 2024, organizations lost over $40 billion just in the U.S. associated with lost productivity and system outages (Blanton 2024). These attacks are predicted to become more likely in the medical field. For example, hackers could exploit vulnerabilities, locking up access to critical systems, whether it be on the medical device manufacturing line or with patient care. This could lead to a halt in manufacturing medical devices, disruption in treatment or care, or even patient death. Another concern that could arise early in the manufacturing process would be if an attacker were able to insert malicious code or compromised components into devices. This could go undetected during production, and before deploying tampered devices to healthcare professionals.

At the other end of the spectrum would be when a device has reached the end of life and is explanted from the patient, but without executing proper protection procedures for both patient information as well as the device. Accordingly, processes



W. Harding et al., *Securing a Connected Future*,
https://doi.org/10.1007/978-3-032-07309-9_4

should be put in place to wipe the data from a decommissioned device to ensure that patient-sensitive data cannot be used by cybercriminals.

As AI and ML solutions continue to increase in demand, potential misuse represents a new risk/threat. For example, AI models could be utilized to manipulate device behaviors, leading to improper treatment and poor outcomes for patients. With highly sophisticated attackers, these incidents could go undetected for long periods of time, allowing the compromised devices to continue collecting patient-sensitive data. Further, with AI and ML being relatively new, regulations need to continue to evolve to ensure medical devices are being designed with cybersecurity in mind. With that point in mind, a former U.S. president tasked the Department of Health and Human Services to establish policy around contemporary health care, which includes the use of AI. That point emphasizes that the use of AI and ML needs to follow the Health Insurance Portability and Accountability Act (HIPAA) and General Data Protection Regulation (GDPR) standards/regulations to protect personal health information.. And with the fast-paced evolution of AI-ML regulations, there is a need to be adaptive and continuously updated to enable medical device manufacturers to stay ahead of potential attacks.

Beyond consideration of hardware vulnerabilities, there is a need to also focus on medical device software vulnerabilities. For example, most medical devices rely on some form of software either directly or indirectly (i.e., during the build process), and as devices become more complex, so does the software.

## *1.2 Cybersecurity Lifecycle—Software as a Medical Device*

With the advancements in medical care, considerations need to be made throughout the lifecycle of a medical device as it specifically relates to cybersecurity. Leveraging software as a platform, the medical device lifecycle can be evaluated in terms of a simplification of the phases (e.g., design, develop, test, deploy, and sustain), and it can then be expanded to include designing for cybersecurity, similar to how a medical device company would design for manufacturability and/or design for reliability.

Designing for cybersecurity throughout all phases of a medical device's lifecycle must be included in the concept phase, where potential cybersecurity risks to hardware, software, and communications represent real threats. And during the design phase of a medical device, there must clearly be mitigations against these threats, which also includes ensuring compliance with FDA guidance associated with cybersecurity for medical devices (e.g., ISO/IEC 27001 and IEC 62443). During the software development phase (e.g., code development, integration, and recurrent testing) a focus should be placed on building solutions that emphasize securing data that is used internally to a device as well as communicated between devices or to/from cloud services. Thus, all data must be encrypted regardless of how the data is used. Furthermore, access to data must be examined from the perspective of restricting who can collect that data and who/what can use the data (human, hardware, software). And as partners, contractors, and vendors are engaged in the development phase, decisions must be made regarding whether third-party software could introduce risks into a device. Lastly, medical device software should be designed with patch and update management in mind so that firmware updates can be made without compromising the functionality of the device or the patient data.

During the manufacturing process, supply chain security should be emphasized. And with many manufacturing hardware solutions being designed and built by third-party suppliers, all manufacturing elements must be evaluated from a quality assurance (QA) perspective as well as to ensure that cybersecurity controls are in place within third-party environments. For example, firmware and/or control system software is rarely evaluated as closely as customer-facing software when ensuring cybersecurity compliance. Accordingly, it is critical that firmware/control system software be examined at all manufacturing steps to ensure harmful code is not injected into the device before the device is moved to the distribution phase. And while considering what could get injected into a medical device, it is equally important to consider the risks to the process equipment that negatively impact an organization's infrastructure (e.g., facilities, engineering, financial, maintenance, etc.).

As part of the medical device deployment phase, and when a device is provided to a customer, it is important to ensure that customer/user training is provided, such that awareness of cybersecurity risks is emphasized. That training should include instructions related to how to securely operate and maintain the device, emphasizing the protection of user/customer data and the continued functionality of the device. Likewise, there is a need for device manufacturers to incorporate real-time monitoring of devices to identify unauthorized access attempts or suspicious behavior. And in hospital settings, medical devices should be deployed across a wide range of segmented networks, which will enable hospitals to separate devices from other critical hospital systems, thus minimizing the impact of exploited vulnerabilities.

Building on points associated with monitoring and sustaining devices, it is important that those processes remain in effect for the entire life of the device. And as new threats and vulnerabilities are realized, software patches and updates will need to be pushed out. Additionally, the methods for pushing out patches/updates need to be evaluated to ensure that validated delivery processes are used. Ultimately, the points made directly tie back to the first phases of device ideation and design, which characterizes the cyclic nature of a medical device lifecycle.

To further ensure that medical devices are designed with cybersecurity in mind, manufacturers need to be aware of the standards that need to be followed. For example, as previously mentioned, there is a need to be compliant with FDA and HIPAA regulations associated with protecting patient data. But it is equally important that manufacturers familiarize themselves with the most recent and updated international standards. Those standards cover how to protect data and devices as well as how to identify and respond to both intentional and accidental compromises. Accordingly, one of the most recently published IEEE/UL standards is the IEEE/UL 2933-2024 TIPPSS standard that is being recognized for its forward-looking perspective and aggregation of legacy cybersecurity standards (Medcrypt 2025).

## 2 Overview & Significance of IEEE/UL 2933-2024 TIPPSS Standard

According to Medcrypt (2025), it is important, when considering why and how you will protect your healthcare data and IP that you establish standard and repeatable practices. Otherwise, it will be hard for you to know that you have taken the correct steps and completed the needed actions for plugging vulnerability holes. Further, if you continue to build your business as well as new products without following established cybersecurity standards, you are putting your business and your customers at risk. That said, not all businesses are large enough to have a Chief Information Security Officer (CISO) who would be aware of the latest cybersecurity standards, so the burden is placed on technology leaders. However, as a reader of this book, you are now privy to the latest IEEE/UL standard that enables you to find everything that you need to know about medical device industry standards in one location.

The standard that any medical device organization should have readily available is the IEEE Std 2933-2024/UL 2933:2024. More specifically, it is called the IEEE 2933 Clinical IoT Data and Device Interoperability with TIPPSS—Trust, Identity, Privacy, Protection, Safety, Security standard (IEEE/UL, 2024, Standard no. 2933-2024). This TIPPSS standard is not only the most significant and recently published, but the Working Group also received the winner of the 2024 IEEE Emerging Technology Award. With a team of more than 300 of the highest-skilled international experts, the standard provides an organization with all the details needed to ensure that the latest cybersecurity measures are being followed (Fig. 1).

**Fig. 1** IEEE/UL 2933-2024 TIPPSS Logo

Approved June 6th, 2024, the IEEE/UL 2933-2024 standard establishes the framework with TIPPSS principles (Trust, Identity, Privacy, Protection, Safety, Security) for Clinical Internet of Things (IoT) data and device validation and interoperability (IEEE/UL 2024, Standard no. 2933-2024. This includes wearable clinical IoT and interoperability with healthcare systems, including Electronic Health Records (EHR), Electronic Medical Records (EMR), other clinical IoT devices, in-hospital devices, and future devices and connected healthcare systems (IEEE/UL 2024, Standard no. 2933-2024. In addition, as shown in the following word cloud, the significant topics that the TIPPSS standard addresses are clear, such that if any of the emphasized topics relate to you, then you will need to be familiar with the TIPPSS standard (Fig. 2).

You might ask why the IEEE/UL TIPPSS standard matters to you, and it would be a good question. So, to help answer that question, it would be best to ask yourself a few questions:

- Will my product interact with patients or healthcare professionals?
- Will my product interact with other products/devices (e.g., IoT tech)?
- Do I have any intellectual property (IP) that I want to secure or not expose?
- Do I have processes, methods, formulas, or procedures that could be at risk?
- Do I have anything to lose if my network, computers, mobile devices, or products are compromised?

If you answered "yes" to any of those questions or if you even paused to consider how to answer those questions, then you need to be familiar with the IEEE/UL TIPPSS standard. And the worst case, you need to make sure that you or a highly skilled cybersecurity expert can understand and follow the requirements established in the TIPPSS standard. Further, as you continue to build products or offer services that are targeted toward the medical industry, you need to have the TIPPSS maturity

Fig. 2 IEEE/UL 2933-2024 Word Cloud

model hanging on your office wall so that you can always be reminded of what needs to be done to mitigate risk. For example, see Fig. 9 for a brief overview of what a cybersecurity maturity model might look like and consider how it fits your organization.

## *2.1 Why Use the IEEE/UL 2933-2024 Standard*

On the darknet, an individual's health records are worth considerably more than the individual's credit card. For example, according to the Privacy Affair Dark Web Price Index (Juan 2025), a credit card number can be worth an estimated $125. While medical records could be worth more than $4000. And because your medical records can contain information, such as social security number, date of birth, name, address, medications, and insurance plan details, they can be used in identity theft, drug abuse, and ransomware attacks. Furthermore, since an individual's medical records are persistent, unlike often-changing financial records, those medical records represent a continuous and rich source of data that can be compromised many times. As was stated by one of the IEEE/UL 2933-2024 officers, the reason for the attacks is that it can be financially lucrative.

## *2.2 IEEE & UL Healthcare Security Framework*

As the world drives toward becoming more digital, IEEE has focused on four technology trends that need to be emphasized. One of those trends is universal human values, which aligns with previous statements that the human body is the next digital platform. Similarly, the IEEE/UL 2933-2024 standard embodies Trust, Identity, Privacy, Protection, Safety, and Security (TIPPSS), and it stands out because it does not just focus on cybersecurity risks but also on establishing trust as well as the ethical use of technology. Accordingly, the technology that the IEEE/UL 2933-2024 standard includes is wearable tech, clinical IoT devices, interoperability with healthcare systems (EMRs and EHRs), hospital devices, and future devices that will be consumed by healthcare systems.

**Elements of the TIPPSS Framework**

**Trust**: Establishing reliable and trustworthy connections among devices, allowing only trusted and designated devices, people, or services to have access.

**Identity**: Ensuring that devices and users are correctly identified and authenticated, to establish their trustworthiness, by validating the identity of people, services, and things (i.e., identity and access management).

**Privacy**: Protecting sensitive patient data from unauthorized access.

**Protection**: Implementing measures to safeguard devices from cyber threats and protect them and their users from physical, digital, financial, and reputational harm.

**Safety**: Ensuring that devices operate safely and do not pose risks to patients.

**Security**: Maintaining the overall security of the device, data, and patients.

## *2.3 Security Issues Addressed by the IEEE/UL 2933-2024 Standard*

Many existing standards address elements captured in the IEEE/UL 2933-2024; however, those existing standards do not address what we consider to be end-to-end security. As was echoed by an officer of the standard, any device or application can become the entry point for individuals looking to breach a system. Furthermore, organizations need to follow a standard that can help them measure, verify, and validate the security of their strategies. Whereas other existing standards do not measure the full security stack (i.e., top to bottom), the way that the IEEE/UL 2933-2024 standard does. And with healthcare organizations not having a clear path to implementing and measuring their security strategy compliance with standards and regulations, it is easy to see why there have been so many successful attacks.

## *2.4 Cost to Healthcare Organizations That Do Not Implement IEEE/UL 2933-2024*

When considering the costs of implementing the IEEE/UL 2933-2024 standard, organizations need to consider what they have to lose if they do not implement the standard. For example, it is a high-stakes game to play when you are trying to protect your organization from data breaches. Where medical device manufacturers and third-party service providers must agree/align on the requirements that will be followed as they collaborate on implementing solutions/systems. Thus, lowering implementation costs by reducing the amount of work needed as security measures are applied to medical devices and systems. However, not all healthcare organizations have the funds to fully implement a security strategy as outlined in the IEEE/UL 2933-2024 standards, but they can use less costly tools to get started. For example, healthcare organizations on a tight budget can use the Medical Device and Health IT Joint Security Plan (Standard 2022) provided by the Health Sector Coordinating Council (HSCC) or use the Vendor Management practices to help them mitigate security risks.

## 2.5 TIPPSS Impact on the Medical Device Industry

Many global cybersecurity standards focus on the safety of humans through the lens of finished products/devices at the adoption phase of a device lifecycle (i.e., when the device is activated or starts to be used by the customer/user). And at most, those other standards delve into discussions associated with subassemblies or the small elements that are combined to form the finished product, or they lightly cover a fraction of cybersecurity topics. However, the TIPPSS standard moves the discussion deeper with a view of safety from the discrete components level (i.e., with hardware) and the single function (i.e., within software), and it moves past the point of just sustaining a product to the end-of-life or decommissioning of the product. Furthermore, most standards associated with the safety of humans do not consider that the safety of a device and its data can result in compromised human safety and IP, specifically during each phase of a device lifecycle. The lifecycle phases that TIPPSS focuses on are concept, design, development, adoption, integration, testing, validation, deployment, sustainability, and end of life. Accordingly, those are 10 phases that are not singularly or thoroughly covered by existing cybersecurity standards.

The TIPPSS standard is made up of 6 elements that work together to provide humans, devices, organizations, and systems with a clear methodology for cybersecurity. Specifically, the 6 elements that are embodied by TIPPSS are summarized for clarity as follows:

**Trust**: Authorize only selected people and/or services to have appropriate access

**Identity**: Check the identity of humans, services, and things.

**Privacy**: Make certain that data associated with devices, humans, and systems is kept private.

**Protection**: Safeguarding communications, systems, and devices from attacks.

**Safety**: Ensure the safety of humans, infrastructure, devices, and systems.

**Security**: Support the assured future of humans, data, devices, and systems.

A deeper examination of the 6 TIPPSS elements enables the reader to better understand what is embodied by each element and the capability that the collection represents. Respectively, those elements each incorporate the following:

**Trust & Identity**

- User-Managed Software.
- Authentication, Context, Audit/Accounting.
- Onboarding—Practices, Processes, System Management, Tokens, etc.

**Privacy**

- Data Protection Frameworks.
- Privacy Impact Assessment.
- Adaptive Privacy Solutions.

**Protection**

- Secure Communications.
- End-to-End Encryption.
- Tamper Proofing.
- Event Detection.

**Safety**

- Disruption of Data Flows.
- Disruption of Patient Engagement.
- Inability to use Devices.
- Environmental Disruption.
- Workflow and Business Disruption.
- Loss of IP.
- Compromised User/Network Credentials.

**Security**

- Assure Sensitive Data Communications.
- Secure End-to-End Communications.
- Secure Device Ecosystem.
- Secure Development Principles.
- Vulnerability Management.
- Assert Security Status.
- Event Exception Handling.

The IEEE/UL 2933-2024 TIPPSS standard was designed to focus on the data transformation and device interoperability associated with the Clinical Internet of Things (CIoT). Respectively, the TIPPSS standard seeks to reduce the risks that are relevant during the stages of design, development, sustainability, and decommissioning of interconnected IoT devices and systems. That said, the TIPPSS standard must be aligned with all phases of a device/solution lifecycle, with a goal to aid in the care of patients and ensure improved quality of life.

The TIPPSS standard should be integrated into an organization's methods and practices, where decisions are made that impact apps, systems, devices, sensors, machines, and humans. By extension, the TIPPSS standard should be applied to both cleared and investigational healthcare devices/systems. And with decision-making in mind, the TIPPSS standard should be applied to any device or system that is vulnerable from a perspective of trust, identity, privacy, protection, safety, or security.

Ultimately, the use of the TIPPSS standard provides providers of solutions with a structured methodology and guidance for their development, engineering, operations, and sustainability departments. So, if your organization incorporates or builds solutions that involve hardware, firmware, software, services, data communications, data transformation, or a combination of those elements, then your organization must comply with the TIPPSS standard.

# 3 Medical Devices/Products—Use Cases

During the development of the IEEE/UL 2933-2024 standard, it was important for the team to establish empathy with the needs of the medical device industry. Similar to how you might use empathy to identify customer/user needs and pain points, the team found it necessary to establish relatable use cases. Accordingly, four use cases were created that represent real-world scenarios of both existing and future-forward examples. And though the focus of IEEE/UL 2933-2024 was on the medical device industry, you will recognize the applicability of the standard across many non-medical environments (e.g., automotive, home construction, travel, education, military, etc.).

When considering the presented use cases, you, as the developer/customer/user of technology, should define your use cases, which should be a better fit for your business and lifestyle. Furthermore, the provided use cases should only be used as a guide and should not replace any use cases that align more closely with your specific environments, processes, and practices. For example, the IoT world is vast and is continuing to grow, and we believe that the example use cases cover most customer needs.

## *3.1 Connected Monitoring Device (e.g., Continuous Glucose Monitor [CGM])*

This first use case might be a scenario where you can put yourself in the role of a patient or caregiver. For example, as a patient, imagine you are out hiking or running, lost in thought and admiring the views, but not paying attention to your glucose levels (Fig. 3). In that role, you would have a sensor attached to your skin, paired with your mobile phone, such that your phone is regularly receiving updates from your skin sensor. Further, your phone would be communicating wirelessly to a cloud-based device monitoring system, which in turn would transmit status updates to a remote support individual. Similarly, in the role of a caregiver (e.g., the mother/father) for a child who is off playing with friends, you would represent an additional layer that aligns with the remote support person. In both roles (i.e., patient and caregiver), timely monitoring is critical, where any disruption of sensing or communications might represent life or death.

Now, consider the ramifications if something or someone disrupted the secure flow of data from the sensor to the mobile phone, cloud, or caregiver/remote support. Or what might be the potential risks if the data being transmitted allowed your personal and device data to be shared with everyone on the internet? The negative results of those examples might appear on the surface as not earth-shattering, but both examples might result in the direct and/or indirect compromising of the patient's quality of life and even the functionality of the glucose monitoring system.

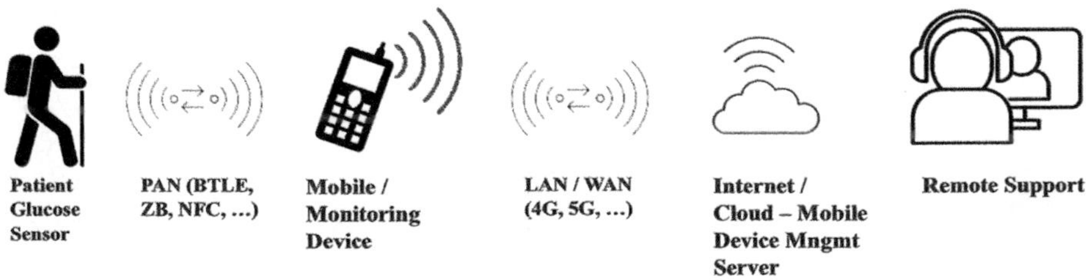


**Fig. 3** Connected monitoring device

Of course, you might ask how disrupted data could impact a patient's quality of life. And the answer is that if your sensor data cannot make it to your mobile device or even to remote support or a caregiver (i.e., as part of an event response), then the patient might experience seizures, cardiac stress, coma, dizziness, or even death, simply because the patient was not made aware of low glucose levels. And if you consider what happens when a diabetic child's data does not make it to a caregiver, you might envision the child collapsing while riding their bike along the side of the road.

All things considered, if the makers of any of the elements embodied by this use case are following the IEEE/UL 2933-2024 standard, then the risks that have been identified will be reduced and possibly eliminated. Thus, by following the standard, system/device manufacturers will construct solutions that cannot have the patient data disrupted while ensuring that data is shared with only those entities who are validated recipients of the data.

## 3.2 *Connected Therapy Device—Connected Automated Insulin Delivery System (AID)*

The second use case moves beyond just monitoring a patient's glucose levels and incorporates an automated insulin delivery device and a physician. And similar to the CGM use case, there is a need to consider data disruption and the sharing of patient data. But in this second use case, we recognize that there is a need to secure the insulin pump device as well as a vial of insulin that is contained within the insulin pump. Further, though the first use case included remote support, we see that the second use case includes a physician who might make patient healthcare decisions based on the correct or incorrect data that is received from the glucose sensor and insulin pump (Fig. 4).

Thinking about the insulin pump, we consider the pump actuator, software, and hardware associated with dispensing insulin, communicating with the sensor, and communicating with a mobile device. And in this case, we must be aware of how the pump software transforms patient and device data so that better decisions can be made that benefit the patient. We must also be aware of how the use of low-quality

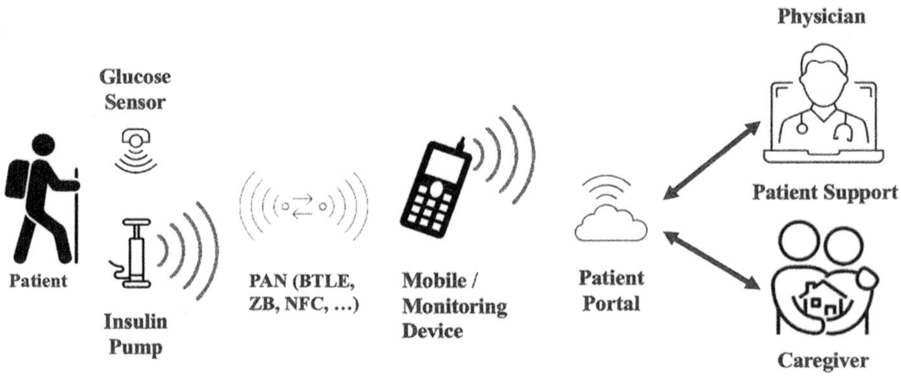


**Fig. 4** Connected therapy device

components (i.e., during the manufacturing of sensing, dispensing, and communications elements) might result in device failure, incorrect insulin dispensing, hacking, and incorrect transmission of data. Furthermore, if there is a disruption of the data between the sensor and pump, then too much or not enough insulin might be delivered to the patient, resulting in seizures and even death.

The added elements of a pump and a physician represent increased risks of intentional and unintentional compromising of a patient's monitoring and dispensing system. Thus, the complexity of the system exemplified by the second use case emphasizes the need for system designers, developers, and manufacturers to consider the steps outlined in the IEEE/UL 2933-2024 standard as those steps relate to hardware, software, and human interoperability as well as the secure transformation of patient/device data.

## 3.3 *Hospital at Home*

In this use case, you can envision that the hospital does not have enough beds and has sent the patient home with technology to monitor the patient's vitals as well as enable communications between the patient, at-home technology, the hospital system, and a healthcare professional (Fig. 5).

With the reduction in cost associated with portable health diagnostic technology, it is very common to find devices, such as a thermometer or a blood pressure monitoring device. However, if you were to visit the home of one of the authors of this book, then you would also find an EKG/ECG, a stethoscope, a glucose testing device, COVID-19 test kits, and even a portable ultrasound device. And though all that technology might sound like too much, the devices offer peace of mind, where the homeowner can use the devices to form a large picture of a patient's current vitals.

Taking a deep dive into the array of home health diagnostics technology, homeowners must stay vigilant regarding the methods by which the home diagnostic

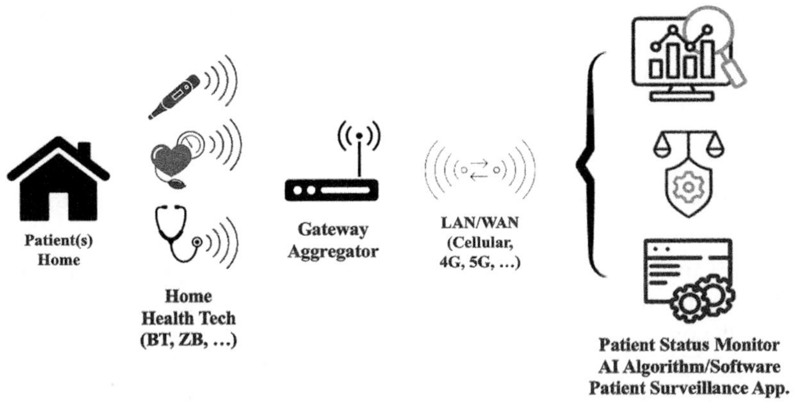


**Fig. 5** Hospital at home system

devices store and transmit patient data. For example, the home diagnostic devices will probably communicate primarily using Bluetooth to a computer or mobile device (i.e., data aggregator). And in turn, that data aggregator will transmit the data as a single file to a remote service that monitors the patient, analyzes the data, and ensures that the data, as well as the communication methods, are secure. Therefore, the homeowner should purchase diagnostic devices and health monitoring services that are TIPPSS (IEEE/UL 2933-2024) compliant, such that all appropriate practices and methods have been applied.

Furthermore, all healthcare monitoring services, including hospital and physician systems, must be validated to ensure that patient data is properly secured, stored, transmitted, and eventually destroyed. That point emphasizes that a healthcare facility or physician that complies with HIPAA does not fully comply with the TIPPSS standard. Due to that lack of complete compliance, patients need to take extra precautions when selecting healthcare services and when deciding to share their health data. When not taking steps to ensure that a patient's data is securely managed, the patient is taking the risk of their data becoming compromised, which could lead to diagnostic data not being properly analyzed and, ultimately, the misdiagnosis of a patient's health. For example, imagine that cardiac data is compromised within a hospital monitoring system and that the wrong medicine is prescribed. Or imagine that an unsecured home data aggregator is hacked, compromising other connected home devices (i.e., IoT tech), such as the doors, climate control, cameras, and home security.

Ultimately, just as a person might check an electric tool for the UL symbol, signifying that the tool is safe to use and meets quality standards, a patient must verify that home health diagnostics devices and services comply with the TIPPSS standard. Patients also have the right to ask if a device or healthcare service provider complies with existing standards. For example, just as a patient might research a doctor to see if that doctor has experience performing a specific procedure, it is important that hospitals, device manufacturers, and healthcare service providers retain documentation and certification material that substantiates their skills and compliance with local, government, and global practices, policies, rules, laws, and international standards.

## 3.4 Home to Hospital

Building off the hospital-at-home use case, it is easy to imagine that a patient has performed at-home diagnostics using mobile healthcare devices, which might identify a health condition that requires returning to a hospital for more intensive testing and therapy. That trip to the hospital might also involve an EMT, ER services, and further clinical care. And in each of those scenarios, the data that the patient first collected at home is critical to providing the patient with expedited care and eventually improved quality of life. Additionally, the hospital should be concerned with the risks associated with allowing a patient's technology to connect to the hospital systems. Specifically, it may be necessary to enable the home diagnostic technology to interface and transfer patient data to EMT technology, to the doctor's office system, and to the hospital system. But how does a patient or caregiver know if they can trust the systems that they will share data with? (Fig. 6)

To help understand the various levels where trust between devices, systems, and individuals needs to be considered, the following graphic is offered as it relates to both the micro and macro views of data exchanges (Fig. 7):

In the figure above, you can see that awareness needs to be focused on the components that comprise a home IoT diagnostics device, and that equal focus should be on the systems, devices, and humans that will connect, communicate, and collect

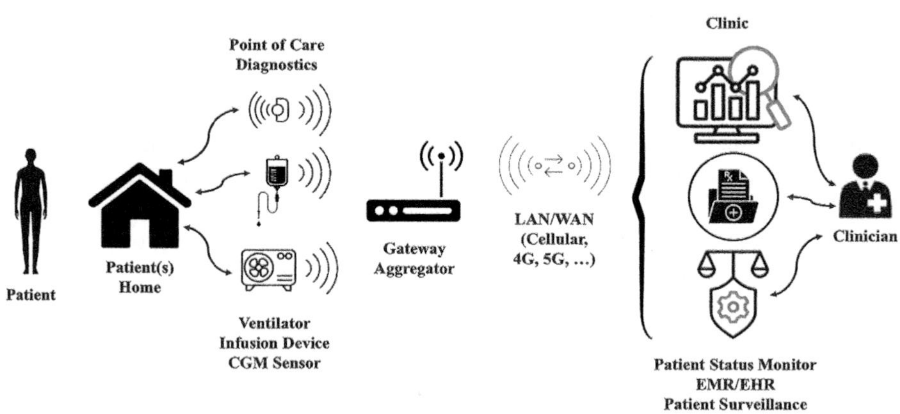


**Fig. 6** Home to hospital system

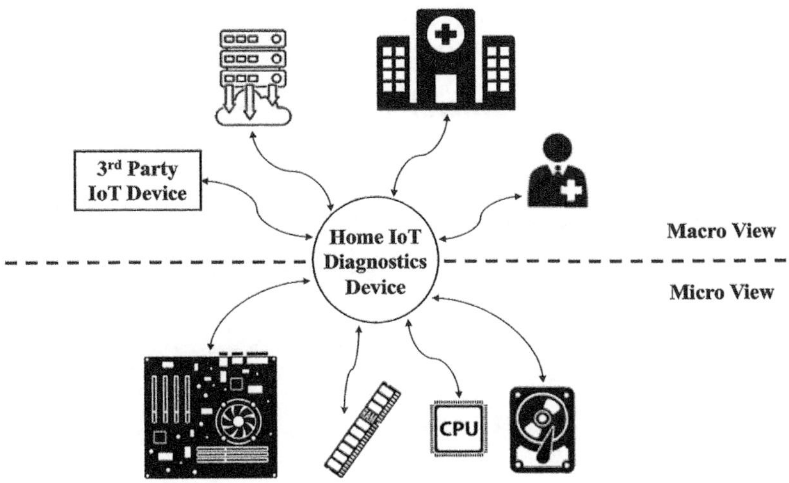


**Fig. 7** Device trust across multiple levels

the patient's data. Consequently, there is a need to establish device trust and identity across multiple levels (i.e., macro and micro).

- Device development and manufacturing.
- Design lifecycle and management.
- Intra-device and inter-device cross-systems trust.
- Data transformation between systems and technology.
- Interoperability with decentralized environments.
- Device-to-human interactions/exchanges.

# 4 TIPPSS Maturity Model

By definition, a maturity model is a framework of security practices, guidelines, and controls that provide an organization with a roadmap for creating effective, compliant cybersecurity and IP protection solutions (Möller and Haas 2022; Rea-Guaman et al. 2017). More simply, a maturity model can be viewed as a tool for establishing a structure that enables one to evaluate current compliance and/or security effectiveness as well as areas for improvement, specifically through the lens of implementing a protection solution. Maturity models are very effective tools when

1. gauging the current cybersecurity state,
2. assessing compliance with the IEEE standards,
3. evaluating exposed risks,
4. identifying areas for improvement,
5. ensuring compliance with requirements (Fig. 8)

It is important to note that a maturity model is not an official or government-mandated validation or requirements document. Thus, it is best to view a maturity

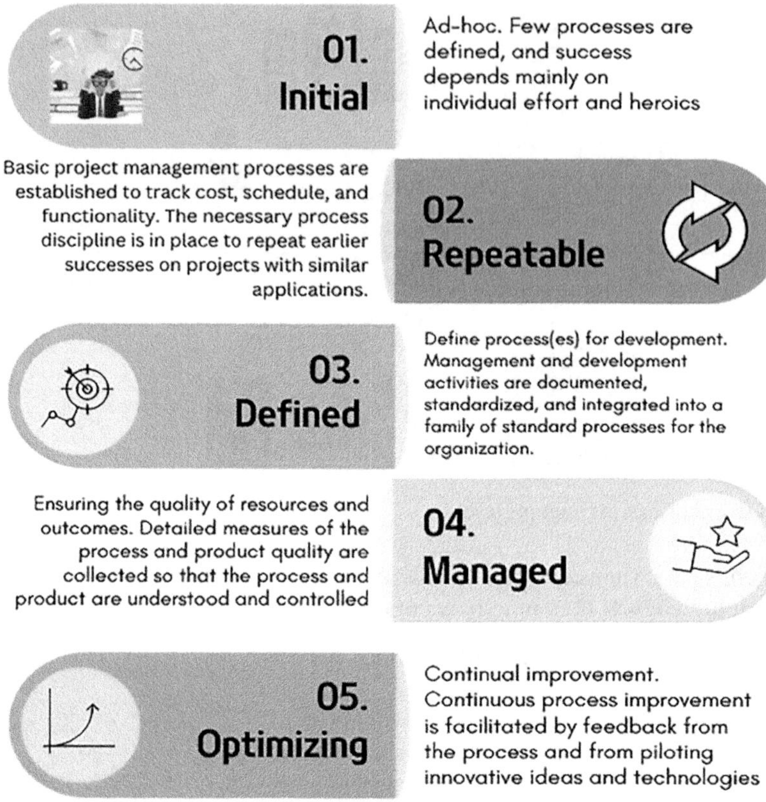


**Fig. 8** Maturity model compliance

model as a guide or maybe a checklist that one uses when seeking to ensure compliance with the IEEE/UL 2933-2024 standard. So, as you examine the IEEE/UL 2933-2024 standard and more specifically the **should** and **shall** statements, the maturity model acts as a guide/checklist that organizations and individuals can use as they seek to reach compliance (e.g., basic, acceptable, and full) with the TIPPSS standard. And to help you determine which level you might be at as it relates to being compliant with the TIPPSS standard, use the following list:

1. Basic Processes/Practices Performed.
2. Processes/Practices are Documented.
3. Processes/Practices are Maintained.
4. Processes/Practices are Reviewed and Reiterative.
5. Processes/Practices are Continuously Optimized.

Building on the levels that serve as a framework for determining compliance with the IEEE/UL 2933-2024 TIPPSS standard, it is suggested that an organization

**IEEE 2933-2024 Compliance Checklist**

Search Description... | Select Requirement | Select Doc. Section | Select PFT Category

29%

| Item Description | Requirement | Doc Section | PFT Category | Doc Location | Page Num. | Compliant |
|---|---|---|---|---|---|---|
| Process established | Shall | Privacy | Process | Sec 5.1 | 66 | ☑ |
| Encryption is enabled | Shall | Security | Technical | Sec 8.2 | 91 | ☐ |
| User authentication measures active | Should | Trust & Identity | Functional | Sec 4.4.2 | 56 | ☐ |
| Process established | Should | Protection | Functional | Sec 6.2 | 75 | ☐ |
| Action enabled | Can | Safety | Functional | Sec 7.2 | 88 | ☐ |
| Manufacturers shall include trust and identity considerations | Shall | Trust & Identity | Process | Sec 4.2 | 38 | ☑ |
| Manufacturers may identify other components | May | Trust & Identity | Process | Sec 4.2 | 38 | ☑ |
| Solution providers shall train all staff in general security principles and basic security hygiene. | Shall | Security | Process | Sec 8.3.2.1 | 96 | ☑ |
| A CIoT device shall allow the user to disable its non-critical technical alerts. | Shall | Human Factors | Functional | Sec 9.3.1 | 116 | ☐ |

**Fig. 9** Maturity model web app with all elements exposed

**IEEE 2933-2024 Compliance Checklist**

Search Description... | Select Requirement | Select Doc. Section | Select PFT Category

75%

| Item Description | Requirement | Doc Section | PFT Category | Doc Location | Page Num. | Compliant |
|---|---|---|---|---|---|---|
| Process established | Shall | Privacy | Process | Sec 5.1 | 66 | ☑ |
| Manufacturers shall include trust and identity considerations | Shall | Trust & Identity | Process | Sec 4.2 | 38 | ☑ |
| Manufacturers may identify other components | May | Trust & Identity | Process | Sec 4.2 | 38 | ☑ |
| Solution providers shall train all staff in general security principles and basic security hygiene. | Shall | Security | Process | Sec 8.3.2.1 | 96 | ☐ |

**Fig. 10** Maturity model web app with process elements exposed

utilize the automated compliance checklist web interface. For example, the IEEE/UL 2933-2024 team has released a strawman web app that provides customers with a tool to review each shall, should, may, might, and can (see Fig. 9).

Furthermore, the tool enables the user to define just the elements found within the standard that an organization might want to comply with (e.g., PFT Category Process). And as can be seen in Fig. 10, the user has chosen to check off only the items that are marked as a Process (i.e., versus Functional, Technical, or unknown). And as can be seen in Fig. 11, the user can enable just those elements found within the Security subgroup of TIPPSS.

So, as can be seen in the maturity model levels guide and the web app, individuals and companies seeking to become compliant with the IEEE/UL 2933-2024 TIPPSS standard will have at their disposal appropriate tools that will make checking compliance with the standard an easier task (i.e., ease of use and usability). Those tools enable companies to delegate and divide up tasks/actions into groups that could focus on (for example) the shall category or maybe just the Privacy area. The task of compliance is a large undertaking and will require the involvement of all relevant stakeholders (QA, Regulatory, Engineering, Operations, IT, etc.) and the effort should not be taken lightly.

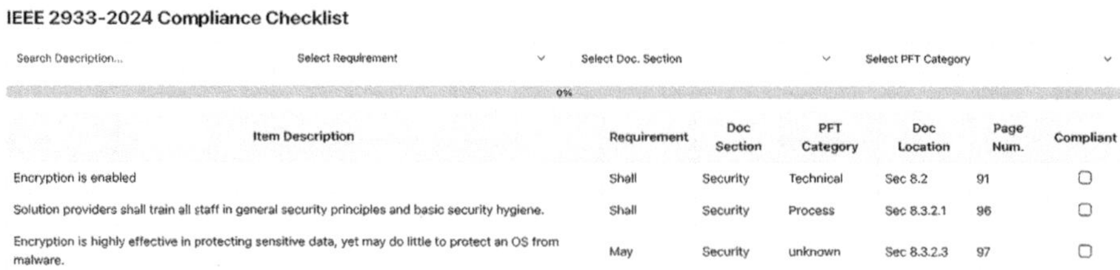

IEEE 2933-2024 Compliance Checklist

Search Description... | Select Requirement | Select Doc. Section | Select PFT Category

0%

| Item Description | Requirement | Doc Section | PFT Category | Doc Location | Page Num. | Compliant |
|---|---|---|---|---|---|---|
| Encryption is enabled | Shall | Security | Technical | Sec 8.2 | 91 | ☐ |
| Solution providers shall train all staff in general security principles and basic security hygiene. | Shall | Security | Process | Sec 8.3.2.1 | 96 | ☐ |
| Encryption is highly effective in protecting sensitive data, yet may do little to protect an OS from malware. | May | Security | unknown | Sec 8.3.2.3 | 97 | ☐ |

**Fig. 11** Maturity model web app with security section exposed

# 5 Neg./Pos. Case Studies Supporting Compliance with IEEE Cybersecurity Standards

The prior information gave you a solid understanding of the IEEE/UL 2933-2024 TIPPSS cybersecurity standard through the lens of its significance, industry impact, use cases, and maturity model. However, what should help you truly understand the importance of following and being compliant with standards is best represented within a series of relevant case studies. Accordingly, we believe that the following case studies represent both the positive and negative consequences associated with following and not following standards.

## *5.1 Negative Case Studies*

### 5.1.1 Case Study #1—Negative Consequences—Company XYZ Data Breach (2013)

**Summary** A large retail store exposed the personal information of over 40 million customers because hackers gained access to the company's network through a vendor's compromised credentials.

***Relevance*** The breach reflected a failure to implement robust IEEE cybersecurity practices specific to access control, network security, and data protection.

- Results <u>Financial Loss:</u> Not including loss due to damaged reputation, company XYZ lost nearly $162 million after insurance.
- <u>Regulatory Fines:</u> Due to the data breach, company XYZ had to deal with lawsuits and penalties associated with several laws, as well as those levied by the Payment Card Industry—Data Security Standards (PCI DSS).
- <u>Reputational Damage:</u> Trust in company XYZ was significantly impacted, which led to a considerable reduction in customer confidence.

### 5.1.2 Case Study #2—Negative Consequences—Credit Organization ABC (2017)

***Summary*** Due to an unpatched vulnerability in their web application framework, Credit Organization ABC exposed the personal information of nearly 147 million people (e.g., social security numbers [SSN], birth dates, and addresses). Additionally, it was discovered that a patch for the vulnerability was available and published months before that data breach, but Credit Organization ABC failed to implement any actions to secure its system.

***Relevance*** It is well known that the failure to apply available patches quickly and the use of outdated monitoring systems reflect poorly on the Credit Organization ABC's capability to apply proactive risk assessments, management, and mitigation.

- Results Financial Loss: Credit Organization ABC paid fines over $700 million, levied by the U.S. Federal Trade Commission (FTC) and other regulatory entities, in addition to money paid to customers as part of legal settlements.
- Impact on Customers: The exposure of SSNs, birth dates, and addresses exposed customers to the risk of identity theft.
- Reputational Damage: Trust in Credit Organization ABC was significantly reduced, which impacted the ability of ABC to conduct normal business.

### 5.1.3 Case Study #3—Negative Consequences—Communications Organization MNO (2021)

***Summary*** Due to the use of unpatched servers, hackers were able to exploit a vulnerability that allowed the distribution of personal information of ~40 million current users and ~60 million prospective customers.

***Relevance*** The exposure of hacked data emphasized the failure of Comm. Organization MNO to implement basic cybersecurity protocols, such as management of patches, which is a primary element of IEEE cybersecurity standards associated with following guidelines and establishing system strength.

- Results Financial Loss: Comm. Organization MNO was required to cover legal costs, pay fines, and compensate customers millions of dollars.
- Impact on Customers: The exposure of customer data, such as SSNs, names, dates of birth, and driver's license numbers, puts those customers at risk for identity fraud and compromised customer safety.
- Reputational Damage: Comm. Organization MNO suffered a considerable lack of customer confidence due to the significant damage to the company's image.

### 5.1.4 General Outcomes Associated with Failure to Comply with IEEE Standards

***Financial Penalties*** When publicly facing companies fail to follow standards associated with data protection laws, they may experience large regulatory fines (e.g., due to GDPR violations).

***Reputational Damage*** When companies fail to protect their customers, trust is one of the first aspects that is lost. Additionally, the loss of customers, investors, and market share is often the result of not protecting customer data.

***Legal Consequences*** In each case of uncontrolled release of customer data, companies will face expensive and time-consuming lawsuits from customers.

***Disruption to Operations*** When a company's vulnerabilities are exposed, there is an inevitable disruption to the company's ability to do business (i.e., downtime). That downtime directly impacts normal operations and customer service.

***Security Debt*** When a company continues not to address issues, the company's security risks can compound over time (i.e., security debt), which will make it increasingly harder for a company to establish a robust and secure system for the future.

### 5.1.5 Negative Consequences Summary

The case studies discussed represent examples of where companies failed to comply with simple and basic IEEE cybersecurity practices and standards. And many of those failures could have been avoided if the companies complied with IEEE cybersecurity standards, such as IEEE/UL 2933-2024. Further, though the financial losses and damage to reputation should be enough to motivate a company to initiate compliance with cybersecurity standards, it is hoped that you will not be one of the victims, since you have been shown the light towards protecting your own company. Lastly, a recent event substantiated by the King's College Hospital NHS Foundation Trust revealed that at least one person died because of a cyber-attack (Carroll 2025).

## *5.2 Positive Case Studies*

The following case studies are offered in support of successes linked to organizations following and complying with IEEE cybersecurity standards. And just as the previous negative case studies provide evidence that there are repercussions associated with not complying with IEEE standards, the following case studies help reinforce that efforts to comply with standards will save an organization money and reputation.

### 5.2.1 Case Study #1—Positive Consequences—Industrial Equipment Organization GHI

***Summary*** Organization GHI is a high-profile organization that is a leader in the industry, which makes the organization a prime target for cybersecurity attacks. Additionally, with Organization GHI connected products being used worldwide (i.e., within their IoT line of devices), any successful attacks would have off-the-charts repercussions. Note: Organization GHI devices are integrated into energy, manufacturing, and transportation industries, which are foundational infrastructures that impact billions of individuals.

***Solution*** Organization GHI adopted communication IEEE cybersecurity protocols and performed risk assessments, such that they could ensure that they complied with IEEE device security for networked devices as it relates to network access control and continuous monitoring of security risks.

***Results*** Following the adoption of IEEE cybersecurity standards, Organization GHI witnessed a considerable reduction in security breaches while improving the hardiness of its industrial systems. Further, through their continuous monitoring and timely patching of systems, Organization GHI was able to detect threats and mitigate those threats, which improved their system uptime.

### 5.2.2 Case Study #2—Positive Consequences—Control Systems Organization DEF

***Summary*** Organization DEF is a controls systems industry leader providing manufacturing systems, such as programmable logic controllers (PLCs), which communicate and control manufacturing equipment and systems that transform and exchange critical data. Those control systems are responsible for maintaining the operational state of technology in the energy, manufacturing, and transportation sectors. Thus, making Organization DEF a prime target for cybersecurity attacks.

***Solution*** With a focus on implementing security within embedded systems and devices, Organization DEF adopted multiple IEEE standards so that they could improve and make their industrial control systems (ICS) more robust. Organization DEF also adopted the IEEE standards 802.1x and 802.11 so that they could ensure that their network access was being monitored and updated on a timely basis. Further, by adhering to and maintaining compliance with IEEE standards, Organization DEF was able to ensure that all its systems utilized the latest encrypted communications channels.

***Results*** Following compliance with existing IEEE standards, Organization DEF improved the robustness of their ICS, which reduced their vulnerability to cybersecurity attacks (e.g., denial of service, ransomware, and internal threats). Furthermore, by adopting existing IEEE standards, Organization DEF improved customer confidence and trust in their systems.

### 5.2.3 Case Study #3—Positive Consequences—Cloud Systems Organization RTS

***Summary*** Organization RTS provides cloud-based solutions related to securely storing sensitive data, performing protected analytics, and the transformation of personal as well as complex industry data. Organization RTS has contracts with government entities. And because of the sensitive nature of data transported across Organization RTS systems, the company is a prime target for cybersecurity attacks.

***Solution*** To reduce successful attacks on their systems, Organization RTS evaluated and adopted IEEE standards, such as IEEE 1686 and 802.1x, specifically as those standards relate to cloud and networked systems/devices. Furthermore, Organization RTS also adopted IEEE standards associated with system design and development practices, which also relate to the secure use of data integrity protocols within cloud systems.

***Results*** Based on the organization's RTS monitoring systems and mitigation practices, there was a marked reduction in successful attacks on the company's cloud platforms. By adopting IEEE security standards, Organization RTS indicates that its ability to provide data protection, encryption, and authentication has improved. Similarly, the adoption of IEEE cybersecurity standards by Organization RTS increased customer confidence in their cloud systems, specifically as it relates to the handling of GDPR and HIPAA-regulated data.

### 5.2.4 Positive Consequences Summary

The case studies discussed represent examples of where companies adopted and complied with IEEE standards, which resulted in companies enhancing the security of their infrastructure. That enhancement effort included regular risk assessments, risk mitigation, and an improved cybersecurity stance. The adoption of IEEE standards stimulated Organization RTS to create a structured framework for integrating best practices associated with securing systems, devices, and networks, while minimizing security breaches while improving trust, and compliance with regulatory agencies. That said, any company that works with sensitive data and is seeking to be a leader in the industry while establishing a trusting relationship with its customers/end users must stay current on the latest IEEE cybersecurity standards.

## 6 Conclusion

The need to follow standards and regulations is clear and extremely important. Not only because it is critical that an end user's data remains protected, but also because devices and software need to remain safe to use. Additionally, by following standards (e.g., IEEE/UL 2933-2024, HIPAA, etc.) and remaining compliant with regulations, Engineers and Software Developers can establish repeatable practices that stimulate a mindset of continuous awareness of cybersecurity risks. Furthermore, awareness of risks and the application of cybersecurity measures helps not only protect customers but also protect an organization's brand, market share, and IP, as well as the systems and technology employed to manufacture safe medical devices.

# Digitally Transforming Manufacturing Operations

## 1 Manufacturing Operations

The computerized (or pre-computerized for that matter) world is in the middle of the next technology evolution, and this time it is round #4. Many different expressions are floating around, such as Digital Transformation, Industry 4.0, Smart Manufacturing, Smart (Digital) Factory, and a few more, essentially pointing towards the same objective: to acquire and use data to drive decisions.

As some of the designations indicate, a major part of the evolution is geared toward manufacturing operations. In the same way AutoCAD and similar programs revolutionized the engineering world in the 1960s, manufacturing facilities will look very different once fully digitally transformed. On closer inspection, the 4th industrial revolution should rather be called the 3.5th revolution, as many manufacturing companies realize they have ignored technology advancements for too long and missed in large part the computerization of equipment and ecosystems, which happened with the third industrial revolution. A catch-up is required, and efforts to transform into data-driven businesses also include said computerization of production systems. While the 4th industrial revolution adds new technologies, such as Additive Manufacturing, Augmented Reality, Artificial Intelligence, and more to the manufacturing floors, in its core, it is connecting production equipment and processes and provides valuable insights that drive decision making (Fig. 1).

Companies feel the pressure of the required transformations as speed is of the essence. The application of digital transformation technologies and concepts gives manufacturers a significant advantage over competitors who might still linger in the past. And here lies the danger of computerized systems that are implemented at speed without the necessary caution. Too often is security, privacy, awareness, and compliance are added to the bottom of the list in a "deploy first, secure later" approach. While this might give companies a sense of being ahead of the game, it can and will incur a multitude of costs and problems at a later stage.



W. Harding et al., *Securing a Connected Future*,
https://doi.org/10.1007/978-3-032-07309-9_5

**Fig. 1** Modern automated manufacturing (Openverse.Org)

While companies must consider, assess, and evaluate new technologies for best fit and high return on investment, manufacturing companies find themselves confronted with another daunting challenge: integrating legacy equipment and systems into this new, high-tech ecosystem. It is much easier to adapt processes for new equipment to conform to required communication and security standards than to create an environment that guarantees consistency and security across all domains. Many aspects must be considered when embarking on the digital transformation journey, with the following 15 aspects being the most important from a Cybersecurity point of view:

- Secure Communication.
- Data Privacy.
- Access Control.
- Device Identity.
- Encryption.
- Network Security.
- Software Updates.
- Incident Response.
- Security by Design.
- Risk Assessment.
- Physical Security.
- Security Policies.

- Security Awareness.
- Compliance.
- Audit Trails.

This section will give an overview of some of the key aspects that must be addressed. It is not a comprehensive guide to secure a company's manufacturing operations when digitally transforming, but it will give insights into the most important attributes attached to the transformation process.

## 2 Secure Communication

Connecting assets and systems to a network to extract data requires a secure transmission architecture. Technologies used for the communication between devices and systems in an IoT ecosystem must be designed to prevent data breaches and unauthorized access. The communication needs to be secure to prevent eavesdropping or data tampering. Companies need to integrate the goal of secure communication as part of their overall Cybersecurity playbook as it is a continuous process that involves monitoring, updating, and improving security measures as threats evolve. Furthermore, employees must be trained regularly about best practices in secure communication as preventing human error is key to avoiding security breaches.

Implementation of Secure Communications—Steps:

- Encryption: All data, whether at rest or in transit, should be encrypted to ensure that intercepted data cannot be understood without the decryption key. Techniques, such as SSL/TLS for data in transit and AES for data at rest, can be used. Especially when dealing with older legacy equipment, it might be a challenge to implement encryption, as some assets do not support this, or re-validation of assets in a controlled environment is difficult. In such cases, data transit can be secured by applying network-level encryption, for example, by using Virtual Private Networks (VPNs) or Transport Layer Security (TLS) tunnels.
- Secure Protocols: Ensure the use of secure communication protocols designed for IoT ecosystems, such as MQTTs, HTTPS, CoAPs, which have security features (i.e., message integrity checks, encryption, etc.) built in.
- Device Authentication: Using technologies such as digital certificates and Public Key Infrastructure (PKI), will ensure secure authentication between devices and systems that communicate over the network. This practice can be achieved through the use of metadata encapsulated token authentication, captured in the patent data driven schema for patient data exchange system (Paffel and Harding 2016).
- Network Segmentation, Firewalls & IDS: Networks should be divided into logical segments, which helps contain breaches as well as prevent the spread of threats, with each segment having its own security policies and controls. For the case when a system has been compromised, it is important to use intrusion detection (IDS) monitoring for suspicious activities and firewalls to control the

network traffic. Monitoring should be a regular task with legacy equipment, and with the help of anomaly detection systems, these tasks can be automated.

- Regular Updates: It is essential that all assets, systems, and software are kept up to date. This will help to protect against known vulnerabilities and contribute to the overall landscape security. When it comes to IoT devices, consider over-the-air (OTA) updates as well as security protocols at message brokers, such as MQTT brokers. Legacy assets might not get patches regularly (or not at all!), therefore, the Cybersecurity playbook should assess the potential threat level and how to deal with this for those assets. When digitally transforming, assets that have been standalone for a long time now become part of the ecosystem and must be treated accordingly.
- Access Control & Physical Security: Access to legacy systems should always be limited to the bare minimum needed. Especially as those systems are not completely redeveloped to meet current security standards, there is a higher risk of unauthorized access through software means literally built in. To mitigate risks, additional authentication mechanisms can be introduced for these systems. This includes virtual or digital as well as physical access to the assets. Again, because those assets most likely lack advanced security features, restricting physical access becomes very important.

## 3 Vendor Systems & Equipment

Incorporating vendor-provided control systems and equipment introduces several layers of complexity to cybersecurity frameworks. The first challenge of establishing a trusted supply chain is often covered by teams in the Global Security Office of an Enterprise. Ensuring that vendors adhere to stringent cybersecurity standards is critical, as any compromise in their systems can cascade into the broader manufacturing ecosystem. This necessitates robust vendor assessment protocols, including thorough background checks, adherence to industry cybersecurity standards (such as ISO/IEC 27001 and IEEE/UL 2933–2024), and ongoing monitoring of vendor security practices. Furthermore, it is essential to conduct detailed security audits of the third-party products before integration, ensuring that they meet the organizational cybersecurity requirements.

Another important puzzle piece to consider is the contractual agreements with vendors. Contracts should include specific clauses that address cybersecurity responsibilities, such as incident reporting protocols, regular security updates, and liability for breaches. It is also essential to include Service Level Agreements (SLAs) that outline the expected performance and security standards the vendor must adhere to.

When integrating vendor control systems and machinery, interoperability and compatibility must be carefully managed. Often, these systems may come with proprietary software and communication protocols, which can create potential entry points for cyber attackers. It is necessary to implement security measures, such as encryption, secure coding practices, and regular vulnerability assessments, to protect these integration points. Good practice would be if a company has developed its standards and protocols and holds vendors responsible for adhering to those. Additionally, network segmentation should be employed to isolate third-party systems from critical operational networks, limiting the potential impact of a breach.

Vendor risk management also extends to the physical security of the equipment. Ensuring that vendor-supplied machinery and control systems are physically secure within the manufacturing environment is paramount. This includes controlling access to sensitive areas, securing devices against tampering, and implementing environmental controls to protect equipment from physical threats, such as fire or flooding.

Furthermore, the integration of vendor control systems and equipment necessitates a comprehensive approach to data security. Data generated and processed by these systems should be protected through encryption, secure storage, and access controls. Organizations should also have clear data governance policies in place, outlining how data is collected, used, and shared with third-party vendors.

## 4 Legacy Systems

Integrating legacy systems into a digitally transformed ecosystem is a huge challenge for all manufacturing companies. Although discussed in the secure communication section, there are more aspects that need consideration when it comes to, simply put, old equipment.

Software does not age as well as wine does; therefore, it often lacks updates, has outdated security measures, and potentially several of the following common vulnerabilities, making them a prime target for attacks:

- Outdated software.
- Lack of encryption.
- Weak authentication.
- Insecure network protocols.
- Lack of monitoring and logging.
- Physical vulnerabilities.
- Hard-coded/no credentials.
- Lack of vendor Support.

For any transformation project in the manufacturing operations space, it is important to acknowledge the potential vulnerabilities that outdated assets can introduce, assess the risks, and prioritize a systematic approach in implementing mitigations.

To address vulnerabilities accordingly, they need to first be identified. Tasks, such as vulnerability assessment, penetration testing, documentation reviews, researching known vulnerabilities for the used technology, and external audits, can help with this. Once identified, the impact must be assessed from an exploitation point of view. "What would the potential impact be if this weakness were exploited?". This should include consideration for data exposure, criticality of system functionality, disruption to business/manufacturing processes, and the potential for creating harmful scenarios. To complete the assessment list the likelihood of an exploit of all the identified issues should be added. If it is very complex to exploit a specific weakness or requires multiple levels of authentication, the likelihood of it being attacked might be relatively low. Bringing these insights together creates an overview that allows the prioritization of which vulnerabilities must be addressed first.

This overview "what" supports developing a remediation "how" and implementation "when" plan that can include system patching, security controls implementation, and replacement for highly critical systems that cannot be secured otherwise. Especially, the "when" is an important factor as the affected systems are running in a production environment, some of which are in use 24/7. Therefore, a solid and agreed implementation plan must include required downtimes for manufacturing lines or the acquisition and installation of new equipment or systems.

And with anything else when it comes to security, this must be a "rinse and repeat" exercise as part of the standard manufacturing and IT process playbook. New vulnerabilities are discovered or created, and addressing those is a continuous process.

## 5 Data Privacy

Data privacy is a critical aspect in the manufacturing industry, especially in the era of digital transformation and Industry 4.0. It is a legal as well as ethical obligation to protect sensitive information, such as patient data, in medical devices. Collected data must only be used for its intended purposes, especially in the context of medical devices where patient data is involved. Any company in this field must comply with regulations such as the General Data Protection Regulation (GDPR) and the Health Insurance Portability and Accountability Act (HIPAA). Handling data privacy is not just a legal requirement but also important for maintaining patient and customer trust.

As companies handle sensitive data, such as customer and patient information, intellectual property, and more, data privacy becomes a key risk that needs to be addressed. The collection of data has significantly increased thanks to technologies like Artificial Intelligence (AI), Cloud computing, and Machine Learning (ML), which often requires companies to rethink their approach to the storage, processing, and therefore safekeeping of the vast amounts of information gathered. Also, to keep costs down and ensure continuous support and maintenance of all the data systems, Cloud providers, such as Amazon Web Services (AWS), Microsoft Azure, or Google Cloud, provide a broad range of services to deal with these requirements. However, this does not release companies from their duty to ensure all sensitive information is safe from cyberattacks. Companies must continue to invest in cybersecurity skills and implement strong data protection controls (Fig. 2).

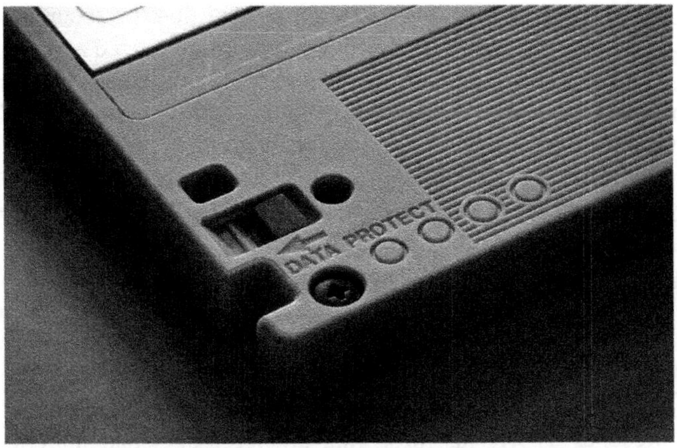


**Fig. 2** Manual data protection (Openverse.Org)

This is just as important as the adherence to any law that is aimed at protecting customers and patient-sensitive data. Regulations, such as the European Union's GDPR, HIPAA, and the California Consumer Privacy Act (CCPA), offer general protection and guidance, but at the same time increase the compliance risk for companies. It requires a very good understanding of the regulatory requirements as well as solid data governance to help create and implement strict security guidelines and measures, required access controls, and up-to-date privacy policies. The deployment of automated data governance systems (i.e., data catalogs) can support a company in classifying sensitive data, enforcing privacy rules, and informing leaders about potential rule infringements.

A fundamental data protection and privacy framework is not only important to safeguard sensitive information, but also to enhance a company's business resilience, build and strengthen confidence of shareholders, and ensure the effectiveness of any privacy policy within the company. Products become increasingly tailored towards very specific market needs, and compliance with privacy and protection requirements must be a core component of any manufacturing process. And with the expected evolution of technology as well as the development of new laws and regulations, data leaders must continue to expand and evolve their policies and actively engage in discussions and assessments on both the technology and regulations side of data protection and privacy.

The following list is a brief overview of several key aspects of how to ensure data privacy in complex manufacturing ecosystems:

- Data Minimization: To reduce the amount of data that is potentially exposed in a security breach, only data that is necessary should be collected.
- Access Control: Restricting physical and digital access to only authorized individuals will help limit the exposure of sensitive data.
- Encryption: Data in transit and at rest should always be encrypted. Data that is intercepted or accessed unauthorized cannot be used without the decryption key.

- Secure Data Disposal: Physical and digital data must be securely disposed of once it is no longer required. This will prevent unauthorized access.
- Privacy by Design: Especially, new manufacturing processes should implement data protection and privacy principles from the start. The use of technologies that incorporate privacy concepts and allow the easy implementation of the company's privacy rules.
- Regular Audits: Internal and external audits should be conducted regularly to ensure adherence to both company and regulatory policies.
- Training: Employees should always be up to date with the latest data protection and privacy policies within the company. Regular and monitored training will help to reduce (and hopefully prevent) breaches caused by human error.
- Incident Response Plan: This should be part of any good data protection and privacy framework and includes guidelines for identifying, containing, and preventing breaches as well as informing all necessary stakeholders and regulatory bodies.
- Compliance with Regulations: Ensuring compliance with existing regulations supports any internal policies and rules. It also allows companies to stay informed about any new laws and regulations.
- Transparency: The collection and use of any data should be handled transparently. Any necessary stakeholders should be kept informed about the data use and how the company is safeguarding and protecting it.
- Security by Design: Thinking about security in the context of the transformation of complex manufacturing ecosystems, Cybersecurity introduces a principle called Security (or Secure) by Design. It refers to the importance of building security into digital products and systems from the very beginning. At the start of a product development lifecycle, cybersecurity goals should be aligned across the organization, be security-focused, and proactive. The principle aims to help protect new technology products or technology-based equipment from malicious attackers gaining access to systems and devices, data (any type), or the entire infrastructure. Security must be part of the development process and not considered on an as-needed basis. Often, associated costs are a factor for companies delaying or minimizing efforts to build security into any design process, which can have disastrous consequences in case a breach or an attack takes place.

Throughout the product lifecycle, it is recommended to conduct regular risk assessments to support the identification of potential vulnerabilities, learn of new exploits and threats, and assess any impact these might have. To counter any findings, Cybersecurity teams can then, based on the risk assessment, create plans including updating patch cycles, strategies to boost security controls, and, if necessary, redesign certain aspects of the security framework.

There are a few more parts to the overall concept of Security by Design, such as the principle of least privilege, which enforces rules around assets only having the minimum required privileges. For example, a component that only needs to read data from a database should never have write access. Systems should also be in their most secure configuration by default. Any changes to security levels must be made consciously. As mentioned before, all systems should be kept up to date with patches and updates.

Security must be considered throughout the entire lifecycle of a product, system, or asset, from its initial design and development, through the deployment and maintenance, to eventual decommissioning. This means that security considerations should not stop once a product is deployed. Regular security reviews, updates, and patches should continue throughout the product life cycle.

The important security aspects and concepts when transforming complex manufacturing operations include continuous risk assessments, which must include the identification of potential threats to assets, the evaluation of vulnerabilities, the assessment of the likelihood of a breach or attack and its impact, as well as the prioritization of identified risks and a solid mitigation strategy. Best practices to secure any system or asset include regular employee security awareness training with the active promotion of a security culture. If feasible, the Cybersecurity team should consider attack simulation strategies to better prepare policies and rules for the eventual case of a breach. Lastly, staying compliant and auditing on a regular basis will strengthen the company's security landscape. An understanding of any compliance requirements is essential, with the necessary control implementation and sufficient documentation. This will assist with any internal or external audit, which should assess policies, rules, documents, storage and handling of data, training efforts and records, and more.

Most importantly, companies should not try to save costs and take shortcuts when it comes to security. Investments in the right skill sets and systems are imperative to best protect against any harmful events, just as fostering a culture of safety and security is.

## 6 Conclusion

After examining the cybersecurity challenges that arise during the digital transformation of manufacturing operations, where legacy systems meet modern connectivity and automation, we can now turn our attention to another critical vulnerability in the medical technology ecosystem: third-party acquisitions. As organizations expand their capabilities through mergers, partnerships, and vendor integrations, they often inherit not only new technologies but also hidden cybersecurity liabilities. In the next chapter, we will explore how these third-party relationships can introduce significant risks, and more importantly, how to identify, assess, and mitigate risks to protect both operational integrity and patient safety.

# Third-Party Acquisitions: Cybersecurity Risks and Mitigation

## 1 Complexities of Third-Party Acquisitions

### *1.1 Third-Party Acquisition Overview*

Acquiring a third-party company, especially a smaller, agile one, is rarely straightforward. These moves often arise from a strategic decision that if you cannot compete with them, buy them. But before finalizing an acquisition, it is critical to grasp why the smaller company might have been able to move faster and capture market share more nimbly in the first place.

Typically, smaller organizations face fewer bureaucratic hurdles and have a leaner management structure, allowing them to pivot quickly, experiment with bold product designs, and shorten time-to-market. The downside is that some smaller companies sacrifice rigorous design and security practices in favor of speed and market penetration, hoping to become an attractive acquisition target for a larger entity. However, once the acquisition is sealed, the larger firm absorbs both the target assets and its liabilities. Accordingly, liabilities may include potential flaws in product quality, incomplete cybersecurity safeguards, or inadequate internal processes. This transfer of responsibility underscores the critical need for robust risk assessments, covering product design integrity, security protocols, and operational discipline, before the deal closes.

Beyond the technological appeal, patents frequently represent a small company's most significant asset. They grant exclusive usage rights, enabling the holder to develop products, block market entry by competitors, and safeguard intellectual property. However, when a large multinational company acquires a smaller firm in for example, in a foreign country, local regulations often protect the rights of local inventors. These protections can prevent the transfer of patent ownership abroad.

Acquired small businesses cannot just be shuttered without incurring substantial fines for patent relocation, which may prove financially unfeasible. Closing a



W. Harding et al., *Securing a Connected Future*,
https://doi.org/10.1007/978-3-032-07309-9_6

company altogether would also forfeit the competitive advantage those patents provide. As a result, a larger corporation typically must keep the local business operational and invest additional resources to maintain and leverage the patents within the confines of a foreign country. At the same time, the acquiring firm must also consider how the small company managed its resources, where it cut corners, and the potential technical debt hidden beneath a rapidly developed product.

Though post-acquisition integration challenges loom large, most often, the acquiring company needs a prompt return on investment, which can lead to swift and sometimes disruptive restructurings, such as laying off professionals who took years to train. Consequently, performing thorough due diligence is not about confirming startup technical or financial viability; it also involves planning for cultural, organizational, and operational alignment following an acquisition.

## *1.2 Essential Pre-acquisition Steps*

**Establish Key Stakeholder Relationships** Effective management of business transitions demands solid relationships with stakeholders in IT security, legal, procurement, compliance, privacy, finance, and supply chain. Strong alignment and open communication streamlines decision-making, which enables you to address emerging risks promptly.

**Maintain a Holistic View of Third-Party Risk** Once you have assessed the criticality of the product or service in question, it is essential to evaluate broader risks the relationship may introduce:

- Financial: Investigate financial stability and solvency.
- Operations: Assess their capacity for operational resilience, including workforce stability and infrastructure readiness.
- Geopolitical/Concentration: Factor in offshoring, geopolitical tensions, and potential disruptions due to natural disasters or regional issues.
- Cybersecurity and Data Privacy: Confirm their protocols for threat and vulnerability management, data protection, and response to emerging cyber risks.
- Compliance and Sanctions: Verify their adherence to local and international regulations.
- Reputation: Check for any industry or media red flags that could tarnish your brand by association.

**Build an Extended Supply Chain Inventory** Assess your entire supply chain, both your existing network and the newly acquired one. Catalog vendors, their dependencies, and potential vulnerabilities. This consolidated view helps you identify and mitigate future cybersecurity risks, ensuring that your operation aligns with your strategic objectives.

## 1.3 Third-Party Risk Management (TPRM)

TPRM is the process used by companies to monitor and manage interactions with all external entities, particularly vendors. This process has been used for years by companies to control the data being transferred in and out of the organization.

## 1.4 TPRM Mergers & Acquisitions Evaluations

Prior to establishing a framework for managing any third-party acquisition, a thorough evaluation must be performed to ensure alignment with your organization's strategic plan.

**Assess Third-Party Needs** Decide if your business requires an additional third party to deliver the same product or service. Compare cybersecurity protocols and third-party suppliers from both organizations to avoid duplicative vendors or gaps in security measures and refine the integration roadmap.

**Conduct Comprehensive Risk Assessments** Before fully integrating the smaller firm, run detailed assessments of their products, infrastructure, and potential liabilities. This process reveals possible security gaps and functional weaknesses that the larger company would need to remediate quickly.

**Plan for Control and Segregation** As you integrate teams, systems, and processes, ensure clarity in roles and responsibilities. Implement access controls, segregate critical assets, and consider additional measures or revised policies if the smaller entity brings heightened security risks.

**Collaborate on Risk Management** For joint ventures or highly integrated acquisitions, consolidate expertise from both entities. Hold regular sessions to align on identified risks, mitigation strategies, and communication plans to keep stakeholders informed.

**Manage Vendors Throughout the Relationship Lifecycle** Continually monitor third-party vendors inherited through acquisition. Enforce contract provisions, ensure performance meets agreed-upon standards, and offboard vendors when they no longer align with your operational or security requirements.

## 1.5 TPRM Divestiture Considerations

When selling a business unit, an established offboarding procedure for third-party vendors is critical:

**Establish Transition Service Agreements** If the divested entity continues to use your organization's infrastructure, define clear agreements about responsibilities, access, and timelines.

**Manage Offboarding Risks** Communicate proactively with vendors to clarify timeframes, responsibilities, and final obligations.

**Deauthorize Access** Immediately revoke physical, data, and IT access. Reset logins, change passwords, and remove any API connections.

**Finalize Contracts and Payments** Review contract clauses, verify all deliverables, and settle outstanding invoices according to termination provisions.

**Address Information Security and Data Privacy** Ensure compliance with your legal and privacy commitments, retrieve or destroy sensitive data, confirm confidentiality requirements, and document the offboarding process thoroughly.

**Update Vendor Management Databases** Maintain complete records of the vendor's history and the reasons for offboarding. This data is invaluable for future risk assessments and potential legal disputes.

**Monitor Post-offboarding Risks** Even after terminating a contract, keep a watchful eye on potential vulnerabilities, such as exposed credentials or compromised data, which could surface later.

**Critical Actions for Managing Risk—Comprehensive Risk Assessments** Identify infrastructure vulnerabilities and plan suitable defenses early on.

**Control Applications and Segregation** Define access privileges and responsibilities, maintaining a clear boundary to protect each organization's interests.

**Collaborate on Risk Management** Foster transparent, ongoing communication about potential risks and ensure all entities understand how to manage them effectively.

**Manage Vendors Throughout the Relationship Lifecycle** Continually assess vendor performance, enforce contract terms, and update your TPRM processes to reflect the joint venture's evolving needs.

## 1.6 Risks Associated with Third-Party Acquisitions

Following the initial evaluation of a potential acquisition in mind, we propose taking a deeper dive into the potential issues that accompany an acquisition. Gaps and/or vulnerabilities within acquired organizations can pose significant problems, and therefore, it is paramount to establish a strategy for the assessment of any existing cybersecurity elements within a potentially acquired entity. Accordingly, the following are examples to consider during cybersecurity assessments of a third party (Fig. 1):

If you are an organization that lacks sufficient cybersecurity experience, it is highly recommended that you engage cybersecurity experts throughout the evaluation, assessment, and audit processes. For example, external cybersecurity firms offer services to conduct in-depth security analysis of an existing company's systems, processes, procedures, and staff. Thus, it is recommended to consider strong contractual agreements outlining the cybersecurity requirements and obligations as a step towards risk mitigation.

With the plan to integrate newly acquired third-party technology and infrastructure into your own company, existing third-party policies, procedures, and assets must be aligned with your cybersecurity framework. Your cybersecurity teams need to employ rigor and a strong focus on eliminating the inheritance of vulnerabilities stemming from the integrated new acquired technology stack, as well as to avoid the creation of new issues. Lastly, to close any knowledge gaps for employees in either company, cybersecurity training or awareness should be an integral part of a successful acquisition.

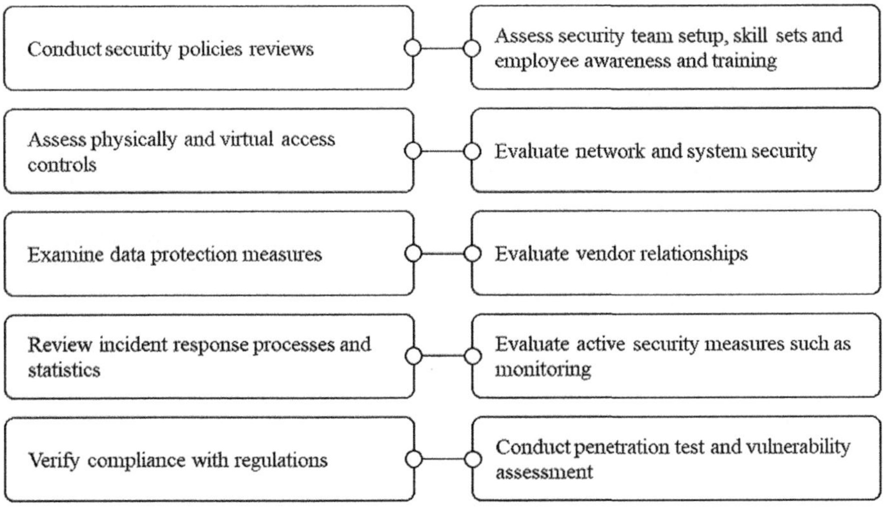


**Fig. 1** Security audit actions

## 2 Analysis of Potential Cybersecurity Threats

Usually, it is business needs that drive decisions when strategically planning a third-party acquisition. The threats that emerge in the information technology domain are often underestimated and might not play a significant enough role in all the considerations taken before finalizing the merger. However, a lack of understanding of what it means to consolidate existing technology stacks can result in considerably higher costs during the merger, or a serious and even arguably existential risk to the company.

The mitigation of risks associated with the third-party acquisition must include comprehensive assessments, the consultation of cybersecurity experts, and an integration strategy ahead of the planned acquisition. Some of the potential mitigating efforts include assessing unpatched or outdated software, aligning processes and protocols, ensuring adequate encryption levels and transmission protocols, reviewing employee access controls, examining security logs and incident responses, as well as reviewing regulatory compliance adherence. It might be necessary to keep entities separated until a strong and secure integration at all levels can be achieved, which should allow for effective safeguarding of digital assets, protect against cyber-attacks, and control the threat of sabotage or unauthorized access. Again, investing in training and upskilling across the workforce is the best weapon against cyber criminals, especially since the majority of all investigated security incidents were due to human error (Marjanov et al. 2023).

During the process of a third-party acquisition, human error is undoubtedly always something that should be considered. It is therefore important to create a mitigation plan to reduce any potential negative impacts. With this mitigation plan in mind, the following list of risks, associated vulnerabilities, and attack vectors is offered as a starting point:

### 2.1 *Data Breaches*

- Risk: Exposure of sensitive information by acquiring a company with a history of data breaches.
- Mitigation: Due diligence: conduct a comprehensive security assessment of the target company's past breaches and current security posture to identify any potential gaps that need to be addressed.

### 2.2 *Dormant Threats*

- Risk: Undetected malware or vulnerabilities within the company infrastructure.
- Mitigation: To identify any hidden vulnerabilities, perform extensive penetration testing and vulnerability scanning before incorporating any acquired systems.

## 2.3 Technology Integration Issues

- Risk: Incompatibility between the acquiring and acquired company's IT systems.
- Mitigation: To ensure a smooth transition and minimal disruption, it is necessary to develop a detailed integration plan that includes compatibility assessments, phased integration, and regular progress reviews.

## 2.4 Insider Threats

- Risk: Employees may pose a security risk.
- Mitigation: To prevent any threats, data theft, or leakage posed by insiders, it is advisable to implement strict access controls, conduct background checks, and monitor for unusual activities.

## 2.5 Regulatory Compliance

- Risk: Non-compliance with data protection regulations can lead to fines and legal issues.
- Mitigation: To ensure both companies are and remain compliant with relevant laws and regulations, conduct regular compliance audits and update policies and procedures as necessary.

## 2.6 Legacy Systems

- Risk: Outdated systems can be vulnerable to attacks.
- Mitigation: To protect against modern threats and to reduce security breaches, it is important to identify vulnerable legacy systems and prioritize the update or replacement of those as part of the integration process.

## 2.7 Cultural Differences

- Risk: Differing security cultures can lead to inconsistent practices.
- Mitigation: To align security practices across the companies, a unified security culture should be fostered through training, leadership support, and clear communication.

## *2.8 Data Privacy Issues*

- Risk: Mishandling of personal data during the acquisition process.
- Mitigation: Effective and up-to-date data privacy policies are imperative to ensure proper data handling practices. Training employees and regular policy reviews will help enforce robust handling and protection of personal data.

## *2.9 Business Continuity*

- Risk: Disruptions during the integration process can impact business operations.
- Mitigation: To minimize disruptions, develop and regularly update a business continuity plan that includes IT and cybersecurity considerations to ensure smooth operations during integration (Fig. 2).

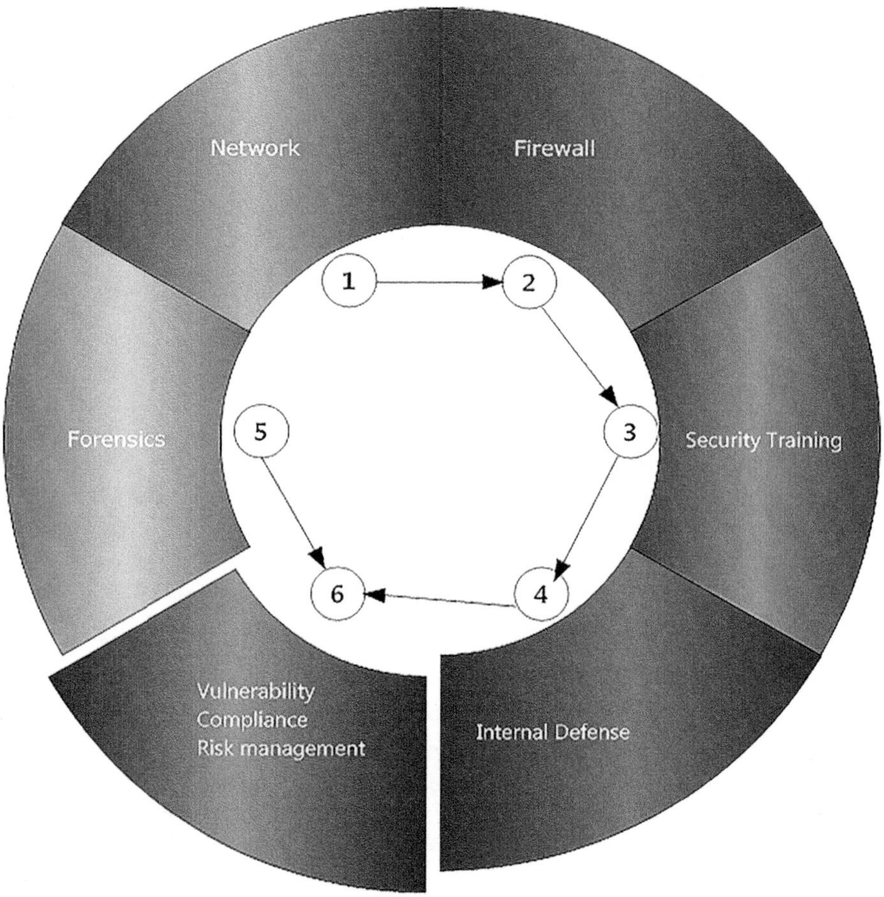


**Fig. 2** Managing cybersecurity risks—6 Layers (Openverse.Org)

# 3 Best Practices, Measures, and Processes for Managing Cybersecurity Risks

As previously outlined, there are many potential risks associated with a third-party acquisition. A non-exhaustive list of victims of cyberattacks includes Target, RSA Security, SolarWinds, and Home Depot, with one of the best-known cases being the data breaches uncovered at Yahoo during its acquisition by Verizon in 2016. Specifically, two breaches have been uncovered going back to 2013 through 2014 with over 1 billion accounts compromised (Thielman 2016; Cheng et al. 2017). With those points in mind, the following list represents measures, practices, and procedures to consider during a third-party acquisition:

- Prepare a strategic and proactive approach.
- Execute comprehensive due diligence assessments and employ third-party security for assessments.
- Create a risk score using Cyber Third-Party Risk Management (C-TPRM) (Keskin et al. 2021).
- Evaluate the cybersecurity posture of the target company.
- Consider insider threat potential and execute thorough vulnerability assessments.
- Consider intentional vs. inadvertent threats (Cheng et al. 2017).
- Review of security policies, procedures, and incident response capabilities.
- Establish strong contractual agreements.
- Integrate ongoing monitoring and auditing.
- Create cybersecurity training and awareness programs.
- Streamline security approach and processes.
- Apply standards and principles.

Supplementing the above measures, practices, and procedures, we offer you the methods to help you further identify risks that third parties might pose:

- Non-Intrusive Methods: These methods are used to assess an organization without any involvement, such as checking the certificates (ISO 27001 and SOC2 Type II Report) to make sure that the organization uses best practices to control their own data and the customers as well and searching the dark web for any breach of confidential data that belongs to the organization.
- Survey-Like Approach: This method involves conducting a survey where a checklist of security controls is asked to be filled out by the organization to evaluate the security level without conducting any technical analysis (done as part of the vendor risk assessment).
- Intrusive Methods: These methods involve direct interaction with the organization's network and require their actions, such as network vulnerability scanning, reviewing logs, audits, and penetration testing reports.

Those measures, practices, procedures, and methods represent a plan that is offered to help you and your organization avoid encountering any negative ramifications associated with third-party engagements. By establishing a tailored risk management plan, you are providing your organization with a unique guide that enhances established cybersecurity best practices.

## 4 Conclusion

Third-party acquisitions have become increasingly common, providing organizations with opportunities for growth and expansion. Acquiring a small, fast-moving company can offer substantial benefits: cutting-edge products, fresh talent, and valuable intellectual property. However, the process that safeguards the successful merger of companies demands a meticulous blend of due diligence, operational alignment, and long-term risk management. You can increase the likelihood of success that will genuinely enhance your company's capabilities while minimizing disruptive aftershocks by taking a holistic approach addressing financial, operational, cybersecurity, regulatory, and reputational aspects.

# AI Risks to IP, IoT, and Product Management

## 1 AI Considerations

With the use of advanced generative AI tools becoming more commonplace, AI-related lawsuits are also on the rise. Some interesting cases to watch include: Doe v. GitHub, Inc. et al., Case No. 3:22-cv-06823 (N.D. Cal. Nov. 3, 2022) a class action lawsuit alleging that GitHub's AI tool for code writing "outputs copyrighted materials without following the terms of the applicable licenses" by training on copyrighted material offered under licenses but failing to abide by the licensing requirements, such as providing attribution; Andersen et al. v. Stability AI Ltd., et al., Case No. 3:23-cv-00201 (N.D. Cal. Jan. 13, 2023) a class action lawsuit alleging copyright infringement based on AI image generation tools trained on copyrighted images; Getty Images (US), Inc. v. Stability AI, Inc., Case No. 1:23-cv-00135 (D. Del. Feb. 3, 2023) alleging that "Stability AI has copied more than 12 million photographs from Getty Images collection, along with the associated captions and metadata, without permission from or compensation to Getty Images, as part of its efforts to build a competing business."

## 2 Identifying Where AI Can Support—Avoid Wasted Investments

AI has become a buzzword, and organizations are deploying AI capabilities everywhere. The problem is that without a clearly defined use case, AI often becomes just another tool in a crowded technology stack, underutilized and disconnected from real business needs. This is like any other technology investment: without proper scoping and integration, AI quickly turns into shelfware, adding to siloed technical debt, increasing IT costs, and frustrating employees. To ensure organizations



W. Harding et al., *Securing a Connected Future*,
https://doi.org/10.1007/978-3-032-07309-9_7

maximize AI potential, it is critical to conduct a thorough assessment of where AI can deliver real value. However, most companies still struggle with this. A McKinsey report (Chui and Malhotra 2018) found that only 17% of companies have mapped out all potential AI opportunities within their organization, and just 18% have a clear strategy for sourcing the data that AI requires. Without this foundational work, AI efforts remain fragmented, failing to scale across business functions or drive meaningful impact.

This problem is particularly acute in cybersecurity, where AI plays a crucial role in threat detection, insider risk mitigation, and IP protection. However, even the most sophisticated AI security tools will fail if they are not effectively adopted. A structured Organizational Change Management (OCM) approach ensures that:

- AI-driven security tools are properly communicated within the organization.
- Employees are enabled and trained to use AI tools effectively.
- Adoption is continuously measured to improve usage and impact.

Without structured adoption, AI risks becoming just another overlooked tool rather than a strategic asset in cybersecurity.

## 3 Methodology for Strategic AI Adoption

To maximize the benefits of AI, organizations should follow a structured methodology that ensures AI supports business goals, security needs, and long-term sustainability.

## 4 Identify the Need

Before implementing AI, organizations must assess the risks to their IP and cybersecurity. This involves:

- Understanding the prevalent cybersecurity and IP theft threats affecting their industry.
- Conducting risk assessments to determine where AI can provide the greatest security benefits.
- Seeking expert guidance to ensure AI aligns with industry-specific risk factors.

Failing to properly define the need and align AI with business objectives can result in wasted investments. According to a RAND study (Ryseff et al. 2024), more than 80% of AI projects fail at twice the rate of traditional IT projects. Specifically, they often fail because they are optimized for the wrong metrics or do not fit into the organization's overall workflow.

## 5 Evaluate AI Vendors and Sustainability

The AI market is saturated with startups, many of which lack the financial stability to provide long-term support. Organizations must:

- Assess vendor viability to ensure the AI provider can survive market shifts and continue delivering updates.
- Verify the AI meets technical certifications and compliance standards.
- Ensure the solution aligns with enterprise security requirements to avoid introducing vulnerabilities.

## 6 Ensure AI Implementation Support

Successful AI adoption depends on strong implementation support. Organizations should:

- Work with AI professional services and solution architects to ensure proper scoping, design, testing, deployment, and hypercare.
- Avoid "cool but useless" AI by clearly defining use cases and integration strategies.
- Ensure that AI deployment aligns with existing security workflows and enterprise infrastructure.

## 7 Measuring AI Adoption and Impact

Once AI is implemented, organizations must measure adoption and success against predefined key performance indicators (KPIs).

- Track adoption metrics to assess how well employees are using AI tools.
- Measure security impact by analyzing whether AI has reduced cyber threats, IP risks, or response times.
- Establish ongoing monitoring to refine AI models based on real-world usage.

Concrete milestones and deliverables should be built into the AI strategy to ensure technology adoption meets expectations and delivers tangible business value.

## 8 Generative AI

While ChatGPT has garnered considerable attention, it is crucial to understand that Generative AI encompasses more than just ChatGPT. For example, within the realm of edge technology and pattern recognition system training, generative AI tools have been used to construct training models that reduce the need for capturing extensive image datasets. This approach has also found application in generating comprehensive datasets, facilitating precise predictions across diverse domains. This technology has the potential to assist organizations in detecting and apprehending malicious individuals attempting to cause harm to an organization.

At the same time, generative AI can present a risk to IP leakage. While phishing emails that appear to come from a co-worker or executive's email address have been around for years and most employees may be aware of this risk, as generative AI gets ever-more sophisticated, so too does its ability to make deep fake videos and phone calls that appear to come from a co-worker or high-level executive at the company, asking for an employee to send confidential documents for an urgent matter, and not to tell anyone about it. The eighth annual Global Incident Response Threat Report from VMware (2022) found that "[t]wo out of three respondents in our report saw malicious deep fakes used as part of an attack, a 13% increase from last year, with email as the top delivery method" (VMware 2022, para. 2). It is thus important to train employees on how to spot deep fake videos or phone calls. Recent studies have shown that humans are "significantly better than chance but far from perfect at discerning whether an unknown actor's face has been visually manipulated by a deep fake algorithm" (Groh et al. 2024, p. 2). Moreover, training significantly improves an individual's chance of being able to spot a deep fake video: in an experience with 25 deep fake videos and 4 real videos, "researchers found that the overall discernment accuracy is 51%, and media literacy training increases discernment accuracy by 24 percentage points for participants assigned to the training relative to the control group" (Groh et al. 2024, p. 2). Providing employees with tips on how to spot deep fake videos and not to fall for scam calls that sound like a real person will increase the organizational ability to prevent IP from leaking inadvertently from the company. One easy-to-implement measure would be advising employees to directly call the company's IT phone line after receiving a request to send confidential documents in order to verify the request.

To mitigate the risks associated with deep fake videos and audio files, a 2024 Deloitte report suggests, for example, enhancing electronic detection techniques by "[w]atermarking audio files" or using "[p]hotoplethysmography (PPG)," and continuing to use and iteratively enhance "biometric algorithms" and anomaly detection in videos and audio/speech files (Ntalampiras et al. 2014). The World Economic Forum has also reported that deep fake detection techniques continue to evolve; as one example, "deep fake detectors can help determine from biometric features, such as a heartbeat or human voice frequency, whether video or audio content is authentic or not" (Buermann and Perucica 2023, para. 7). As the sophistication of deep fakes continues to evolve, so must an organization's detection and protection techniques. Thus, just as cartographers used to add "fictitious entries" like fake towns or non-existent streets (i.e., trap streets) into their maps to detect if someone had stolen their work and committed copyright infringement (Zhang 2015), so too can modern organizations help detect and prevent makers of deepfakes from facilitating phishing attacks and appropriating the likenesses of their executives to steal intellectual property by watermarking their videos and audio files, for example.

# 9 Protecting IP from AI

The following use cases provide exemplary ways that AI can assist a company's security organization as they relate to cybersecurity and IP protection:

## *9.1 Threat Detection & Response—Using AI*

- Use Case: Tools that detect cybersecurity threats in real-time and prioritize cybersecurity threats.
- Example: Detection of abnormal traffic patterns signaling a potential malware infection or unauthorized login attempts from suspicious IP addresses during non-working hours or system breaches.
- Benefit: Accelerates threat detection and response, reducing the risk of undetected breaches and minimizing potential damage.

## *9.2 Insider Threat Detection—IP Theft Prevention*

- Use Case: AI-based User Behavior Analysis tools analyze employee behavior patterns to detect anomalous activity, such as an employee attempting to copy a large volume of files before leaving the company or accessing or downloading a higher-than-usual number of files, relative to their access and download activity throughout their time at the company.
- Example**:** AI can flag suspicious actions, such as:
    - A sudden increase in file downloads.
    - An employee accessing documents outside their usual work scope or department.
    - Printing a large number of documents before resigning.
- Benefit**:** Detects potential IP theft in real time, preventing trade secret leaks before they happen. Considers that certain employees may be downloading or moving large numbers of files as part of their typical job responsibilities, and tailors what is flagged as unusual download activity for each employee.

## *9.3 AI-Powered Patent & IP Infringement Monitoring*

- Use Case: AI automatically monitors patents, trademarks, and copyrights, alerting organizations to potential infringements by competitors.

- Example: AI scans new patent and trademark filings, product launches, and research papers, identifying similarities with organizational proprietary technology or product names and trademarks. It can also detect if a competitor files a suspiciously similar patent or trademark.
- Benefit: Enables early detection of IP violations, allowing organizations to take proactive legal action to protect their innovations, such as opposing the grant of a similar patent or trademark.

## 10 Artificial Intelligence (AI) Chatbots

We have touched lightly on the topics of ChatGPT and Generative AI, which are fundamentally aligned. But with the advancements in the development and use of chatbots, the topic of AI Chatbots requires a deeper examination. For example, many companies have enabled some form of online automated presence to help customers in the same way that FAQs have been used in the past. However, those automated tools have evolved, and we now find that as companies seek to humanize and create a more friendly customer experience, the use of AI Chatbots has become more prevalent.

But what is an AI Chatbot, and how does using one pose a risk to me as a customer, to the company as the deployer, and how can AI Chatbots be misused to give up intellectual property, trade secrets, or disinformation? The answer will surprise you and might shock you into implementing greater protection measures to ensure that you are not exposing protected information or even granting attackers an easier method to enter your company data infrastructure. Further, you will learn how the use of AI Chatbots might put your company at risk, specifically as it relates to loss of IP, loss of market share, or being liable for the distribution of misinformation (Fig. 1).

But before we jump into a discussion of what an AI Chatbot is, let us establish a small foundation regarding what AI is. To start with, AI can be generalized through several different terms, but the most common embodiments or types of AI are narrow/weak, generative, agentic, reactionary, geospatial, situational, emotional, and true AI. Further, true AI has not arrived yet, but when it does, there is a simple mnemonic that will help you recognize if true AI has arrived. The mnemonic is LUCAFIA (**L**earn, **U**nderstand, **C**reate, **A**dapt, **F**ormulate **I**deas, **A**pply), and it represents the following characteristics of true AI:

- True AI can **l**earn.
- True AI can **u**nderstand information.
- True AI can **c**reate knowledge.
- True AI can **a**dapt to changes.
- True AI can **f**ormulate **i**deas.
- True AI can **a**pply knowledge.

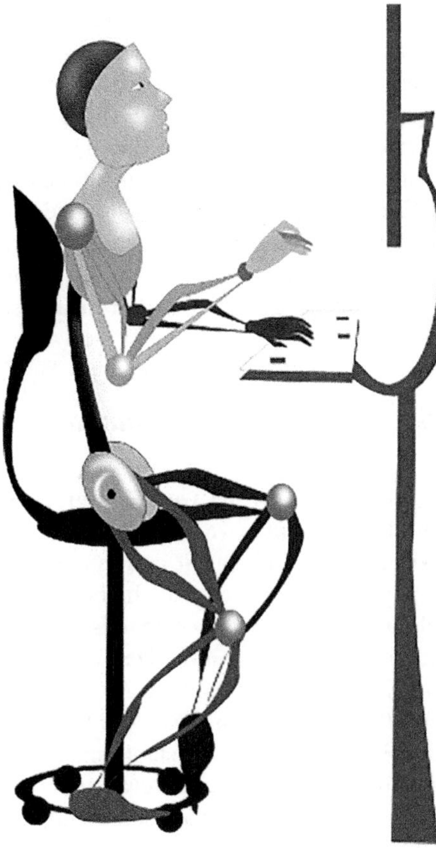

**Fig. 1** AI Chatbot sharing secret information (Openclipart.org)

## *10.1 Common Types of AI*

- *Narrow AI*: Narrow in scope, such as facial recognition, Siri, Alexa, Google Assistant (aka Weak AI) is limited to specific tasks as well as cannot learn.
- *Generative AI:* Is used to create content (also considered weak AI—AI that functions within a set of predefined rules, patterns, programming, and parameters.
- *Agentic AI:* Works autonomously using data derived from LLMs. Unlike generative AI, agentic AI tools do not just respond to requests but will execute actions bounded by established rules. and in most cases, can work independently or without the need for human guidance. For example, Agentic AI is good for use in security systems where immediate detection and a response are needed.
- *Reactionary AI*: Immediate response, no learning (e.g., rule-based or deep search AI).
- *Geospatial AI*: Robotics, navigation, self-driving cars (aka positional AI).
- *Situational AI*: Makes real-time, context-aware decisions (i.e., adapts based on context but still may not learn long-term).

- *Emotional AI*: AI that is more aware of emotions, speech tones, and other relevant human cues. Specifically, emotional AI is more perceptive of subtle human emotions, allowing AI to be more engaged in all psychological aspects of a human. Further, emotional AI provides systems with the ability to react to human emotions collected from verbal and non-verbal cues (e.g., gestures, facial expressions), thus enabling systems to be empathetic to human needs.
- *True AI*: LUCAFIA: Able to make decisions without 100% certainty of the outcome/answer. Applies reasoning, makes reasonable guesses, and possibly speculates on something unknown (i.e., something not previously encountered). Might take risks (e.g., calculated or measured) and is not risk-averse.

## 10.2 What Is an AI Chatbot?

Let us start with what we are familiar with, specifically as it relates to electronic tools that help improve our lives. For example, Siri and Alexa are technically virtual assistants with advanced conversational AI capabilities, but within the scope of what is a bot, they are also chatbots. The AI capabilities relate to using AI tools to help understand voice commands. And though Alexa and Siri are virtual assistants, their role could be summarized as a VoiceBot. Additionally, many people confuse machine learning (ML) with AI, where, if we consider what Siri and Alexa represent, they are tools that use ML to learn from past experiences and improve their capabilities as virtual assistants and can even pick up on human emotional nuances, as well as the contextual relevance between conversations.

Moving from virtual assistants and VoiceBots, we can start to see how those tools differ from AI chatbots. For example, many chatbots employ natural language processing (NLP), natural language understanding (NLU), and natural language generation (NLG) to enhance their capabilities when interacting with humans in either written or voice-based exchanges. But AI chatbots take the effort to the next level using large language models (LLM), which are continuously being trained on large volumes of data associated with human language and knowledge. And through their use of ML, AI chatbots can handle very complex discussions where there might be emotional or conversational variations (e.g., humor, sarcasm, anger, joy, etc.). However, the accuracy and capability of an AI chatbot depend on the accuracy of the data that is found within LLMs. So, if AI chatbots are fed incorrect data, then the answers that an AI chatbot provides may be equally incorrect.

For a company seeking to deploy an AI chatbot securely within a protected organizational environment, the company would need to create a custom LLM with specific company data gathering tools and storage that is not vulnerable to outside access or influence. And the internal AI chatbot system would need to deploy a retrieval augmented generation (RAG) tool that would feed data to the internal LLM. Further, an internal RAG depends on input from users within the organization as well as on being granted access to all forms of data exchange and transformation, which prompts the question: Who ensures the integrity of the data being fed to the AI chatbot? Is it a human or another automated system?

To answer the question related to who or what ensures the integrity of LLM data, here are some steps and actions that should be performed by any company that wants to avoid issues with inaccurate data, such as what might be found in a Wiki page.

- Human Review: With LLMs containing huge amounts of data, it is nearly impossible for humans to check it all, but through random sampling and auditing, humans can review portions of the LLM data, verify the accuracy of that data, and annotate areas of concern.
- Cryptographic Hashing: Similar to embedding a calculated checksum in a unique identifier, cryptographic hashing is where a numerical value is created for a block of data, which can be checked later to determine the integrity of the data (i.e., changes to or corruption of the data).
- Data Validation Practices: Another action that assists the human in verifying the integrity and accuracy of data is when checks for missing data are performed or when the data is checked for inconsistencies as well as outlier data set issues.
- Data Filtering & Pre-Processing: An effort that helps reduce the tasks that humans might perform when checking the integrity of data is to run the data through some pre-processing tools. Those tools help clean the data collected in its raw form before it gets stored in the LLM. Some of the cleaning actions are to look for irrelevant or biased data, which ultimately improves the quality of the data that ends up in the LLM.
- Compliance Checking: In addition to running data through pre-processing tools, it is important that a healthcare organization be compliant with relevant regulations, which might include the need to ensure that data privacy concerns are being addressed. Accordingly, there is a need to check the filtered/pre-processed data to ensure that it aligns with relevant regulations and laws.
- Data Model Monitoring & Evaluation: Lastly, to ensure that the data being fed to AI chatbots are accurate and of the highest quality, it is important to regularly monitor and evaluate AI chatbot output data, checking for inaccuracies and potential biases. For example, there was a case where an HR chatbot was biased in favor of males since the only data being fed to the chatbot related to males. So, if a female was evaluated by the chatbot for a job, it was highly probable that the female would be rejected, since the female did not align with the male model.

## 10.3 What Are Some Risks Associated with Using an AI Chatbot?

The primary risks that are associated with using or misusing AI-enabled chatbots can be summarized as data leaks, system breaches, improper collection and storage of data, lack of methods for destroying data, lack of regulatory compliance, and openly sharing confidential data publicly. For example, humans must keep in mind that an AI chatbot derives its data from LLMs, which could inadvertently be fed confidential information (i.e., IP, financials, business strategies, personal information, etc.) by employees being careless with that data. That feeding of data into an LLM sounds intentional, but when a human poses a query to a chatbot or even posts questions or general information in open forums, then the human has inadvertently fed an LLM that can be used by an AI chatbot. And though there are public, internal, and enterprise (i.e., licensed) chatbots, the same risks exist when data is not properly protected. Accordingly, the risks associated with exposing IP to AI chatbots are as follows:

- Accidental Disclosure**:** Even when humans believe they are being careful with IP-related data, the questions or prompts that an AI chatbot initiates could lead the human to inadvertently share confidential information. Furthermore, humans rarely know the constraints within which an AI chatbot works, where there might be assumptions that it is safe to openly converse with an AI chatbot.
- Misuse of Ideas**:** Building on the point of accidentally disclosing IP data to an AI chatbot, many humans might think nothing of asking an AI chatbot questions related to ideas and concepts that the human is working on. And since an AI chatbot collects data that can be used to enhance the LLMs that feed it data, then there is a good possibility that the information that the human provides in a question could result in that information being exposed in future answers, because of other humans asking similar questions. Ultimately, the exposure of IP could infringe on human rights.
- Training Data**:** As stated in the misuse of ideas, AI chatbots can expose protected information. And since AI chatbots are trained using vast amounts of data, the training material could include third-party protected data (e.g., patents, trade secrets, copyrighted material, etc.) along with the data provided by humans interacting with the AI chatbot. In other words, if a human inadvertently exposed sensitive data to an AI chatbot, then that data becomes training data that the AI chatbot might make available to other humans and systems.
- Data Collection and Storage**:** Regardless of the type of AI chatbot that a human might interact with, there exists a risk that the data might not be securely stored. And if the information shared is confidential data that is not properly stored and protected by the AI chatbot, then there exists a risk that AI chatbot data storage might be vulnerable to unauthorized access by others looking to misuse the data collected by the AI chatbot.

## 10.4 How to Mitigate IP Risks When Using AI Chatbots

One can never assume that interactions with an AI chatbot will be protected and kept confidential. Training, awareness, and attention to details are key practices that will help a human from accidentally exposing healthcare company IP and personal information. And if a healthcare company wants to maintain a market share and protect its IP then the healthcare company needs to be proactive when mitigating the risks of using AI chatbots. Therefore, a healthcare company must train its employees and employ systems to ensure compliance with security best practices. That said, the following five practices represent a recursive methodology that will help a healthcare company reduce the exposure of confidential data.

- Be Careful with the Information that is Shared: Never give away the secret recipe. In other words, when interacting with an AI chatbot, never expose company-sensitive information, formulas, practices, methods, business plans, company strategies, and generally any IP with an AI chatbot.
- Review AI Chatbot Privacy Policies: Read the privacy policies of any AI chatbot that you plan to interact with. For example, become familiar with how data will be collected, where your data will be stored, and how your data might be used.
- Use Internal Chatbots: Never rely on a public AI chatbot for your employees to interact with. Specifically, if a healthcare company wants to deploy an AI chatbot, then it is best to use an internal AI chatbot that is designed with the company's needs in mind and where the AI chatbot is configured to protect data used in its training.
- Monitor Chatbot Interactions: All interactions with an AI chatbot must be regularly monitored and validated. It does not matter if humans or automated security systems are used; data exchanged with an AI chatbot must be continuously checked for leaks of company-sensitive data.
- Consult Legal Counsel: Lastly, when in doubt, the healthcare company's legal team needs to be kept in the loop regarding AI policies, employee interactions, and potential vulnerabilities associated with the AI chatbot and the storage of associated data. Further, when it comes to interacting with an AI chatbot, the company legal team must be consulted regarding methods for handling company sensitive information and how that information might be used both internally and externally to the company.

## 10.5 Lost Market Share—Poor User Experience, Lack of Personalization, & Over-Automation

The world may never move away from wanting a person-to-person experience, that is until a true AI tool is created. When we think about why interaction with a human is critically needed, we can consider how true AI differs from weak or narrow AI tools, such as AI chatbots. Specifically, as it relates to empathizing with user needs and the use of human intuition to help resolve an issue or answer a question (e.g., the healthcare industry). Accordingly, a move to using only AI chatbots can cost company customers who are seeking personal attention and human interactions. Note: It is important to remember that humans are social animals who thrive because of their interactions with other humans.

For example, in research and search engines as an industry tool, dependence on AI chatbots would result in reduced reliance on search engines. Additionally, consider if the movie/TV industry created only AI-based content, and with a world obsessed with celebrities, it is almost unimaginable to think that the amazing uniqueness of your favorite actor would be lost. That point could also be applied to the publishing industry, where the industry would collapse if it only generated content that was being distributed.

To help mitigate the issues associated with the loss of market share due to a dependence on AI chatbots. It might be a good idea to try a hybrid approach that combines manual, semi-automated, and automated practices and processes. Further, though it is more costly to pay for humans (versus only using AI chatbots), it is suggested that human or person-to-person interactions always be an option for industries that depend on providing customers with a social experience. And with humans having a dependency on other humans, there should never be a case where a company eliminates the humans that provide customers/users/patients with personalized attention.

## 11 Liability Associated with Distributing Misinformation

Thinking about AI in more general terms, the United States provides its people with a certain amount of protection associated with freedom of speech (i.e., a constitutional right). However, humans are not afforded that same protection in other countries. For example, there are cases where news agencies and entertainment companies have purposely misled their customers into believing what they want them to believe, such as shaping the narrative of a program associated with true crimes to fit the company's desired goals of suggesting that an innocent person is guilty (i.e., without due process). Thus, motivating their customers to come to the wrong conclusion, which might result in disparaging an individual's credibility or making it appear that a business is out to hurt people. Accordingly, the use of misinformation to solicit a reaction from the public can have very dramatic results.

Misinformation in the public eye is one obvious area where a company can be and should be held liable for distribution, but another important area where misinformation can result in negative consequences is research. For example, if a researcher falsely claims (without substantiated evidence) that certain vaccines can lead to autism in children, the result of that misinformation might mean that many children do not receive needed vaccines that keep them from contracting, for example, polio. And though the researcher is ultimately liable for distributing that misinformation, the damage that results from that misinformation can exist long after the researcher has been held accountable. Note: Speculative research published through a reputable publisher could result in the publisher being held liable, paying large fines, and losing credibility.

Furthermore, since many researchers rely on search engines and research tools to help dig through the mountain of research data, it is easy to see how misinformation provided by an AI chatbot can result in the creation of inaccurate or speculative results. And considering that most AI chatbots derive their data from many internet based databases, it is easy to see how an AI chatbot is only as good as the data that is used to train it. For example, information sources, such as Wiki pages are filled with inaccurate information, since there is no official process for validating and authenticating data entered on a Wiki page. And if a Wiki page is biased against a particular science, race, or methodology, the results provided by an AI chatbot would also be influenced by the associated misinformation.

# 12 Unique Cybersecurity Risks—Robot Surgeries & AI-Driven Triage Systems

Of course, AI can and is used for the good of patients every day. For example, the integration of artificial intelligence (AI) into patient diagnostics and surgical planning is revolutionizing healthcare. AI is being utilized to analyze imaging from X-rays, CT scans, and other outputs, allowing healthcare professionals to gain insights that are often missed by the human eye. Additionally, medical devices like robots used in spinal surgeries and diagnostic tools like the PillCam are transforming patient care, enabling more precise diagnostics and treatment planning.

AI-based systems like Aidoc hold value in the Radiology, Cardiology, Neurovascular, and Vascular fields, where they help prioritize patients with high-risk conditions, improving response times. They can also assist with complex surgeries, such as spinal procedures, by enhancing the accuracy of preoperative planning and image enhancement.

To bring this vision to life, the future of cloud computing will further enhance AI potential by offering greater computational power, improving image quality, and enabling faster, more precise diagnostics. However, today we still heavily rely on existing hardware within hospitals, often without cloud connectivity. While this reliance poses limitations, the shift toward cloud integration will unlock more advanced capabilities. Nonetheless, like any connected medical system, the integration of AI into diagnostics and surgery introduces distinct cybersecurity risks.

There are of course cybersecurity risks posed by AI-driven medical systems and devices, particularly in relation to GDPR and HIPAA compliance. Where it is necessary to discuss mitigation strategies, associated with medical imaging, internet gateways, cloud-based solutions, all with an emphasis on securing remote connectivity, diagnostics, and system updates.

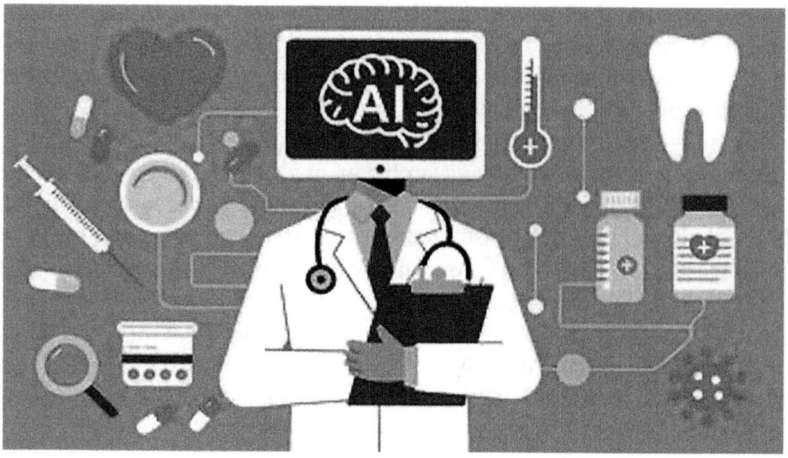


**Fig. 2** AI diagnostics technology (Free-Vectors.Net)

## *12.1 Why Use AI-Based Systems for Patient Diagnostics? (Fig. 2)*

**Enhancing Diagnostic Accuracy** AI-based systems improve diagnostic accuracy by analyzing radiology images like X-rays and CT scans. In many cases, these images are of suboptimal quality, making it difficult for doctors and clinicians to assess the patient's condition accurately. The declining ratio of medical professionals to patients creates even more delays as they are forced to analyze a higher volume of imaging reports. This can result in an increase of mistakes and an increase in the amount of time to get results to a patient in critical condition.. AI, however, can process and enhance these images to detect subtle anomalies like spinal injuries, which can be crucial for surgery planning.

For spinal surgeries, AI can assist in generating detailed models of the spine, allowing surgeons to plan procedures more precisely. These insights reduce the likelihood of complications and improve surgical outcomes. Images can also be analyzed for diagnosing complex conditions like, pulmonary embolism, or even intestinal disorders using tools like PillCam or Aidoc AI to help clinicians prioritize patients.

**Prioritizing High-Risk Patients** AI-driven diagnostic systems are also invaluable for triaging patients. AI algorithms can quickly scan large volumes of radiology data to identify high-risk patients who require immediate attention. This is particularly beneficial in emergency situations, such as detecting life-threatening conditions like intracranial hemorrhages or pulmonary embolisms. This enables healthcare providers to act faster and improve patient outcomes.

**Cloud Computing for Greater Computational Power** The ability of AI to analyze complex medical images is significantly enhanced by cloud computing. Cloud infrastructure provides the computational resources necessary for advanced AI models to process data quickly and efficiently. In settings where high-resolution images are required for surgical planning, such as in spinal surgeries or radiology-based diagnostics, cloud-based AI can help overcome the limitations of local computing power, ensuring better diagnostic results at a lower cost.

## *12.2 AI-Driven Diagnostics & Cloud Computing Risks*

While AI-based systems offer immense benefits, they also present significant cybersecurity risks, particularly when connected to the cloud. Below, we outline the primary risks and potential consequences:

**Data Breaches and Patient Privacy Violations** The sensitive nature of medical data makes AI-driven diagnostics a prime target for data breaches. If unauthorized entities gain access to the patient information stored in the cloud, it could lead to privacy violations, exposing healthcare providers to fines under regulations, such as GDPR (up to 3% of the annual revenue) (in the European Union) and HIPAA (in the United States).

**Risk Mitigation** To mitigate this risk, healthcare organizations must encrypt data both in transit and at rest. Large companies prefer to delegate responsibility to third-party companies that can provide services, such as a cloud-based medical image exchange platform that is designed to facilitate the secure and efficient exchange, storage, and sharing of medical images (such as X-rays, CT scans, MRIs, etc.) between healthcare providers, physicians, and patients. It supports compliance with healthcare regulations like HIPAA and GDPR while enhancing interoperability between different systems that offer robust encryption protocols, such as: (Fig. 3)

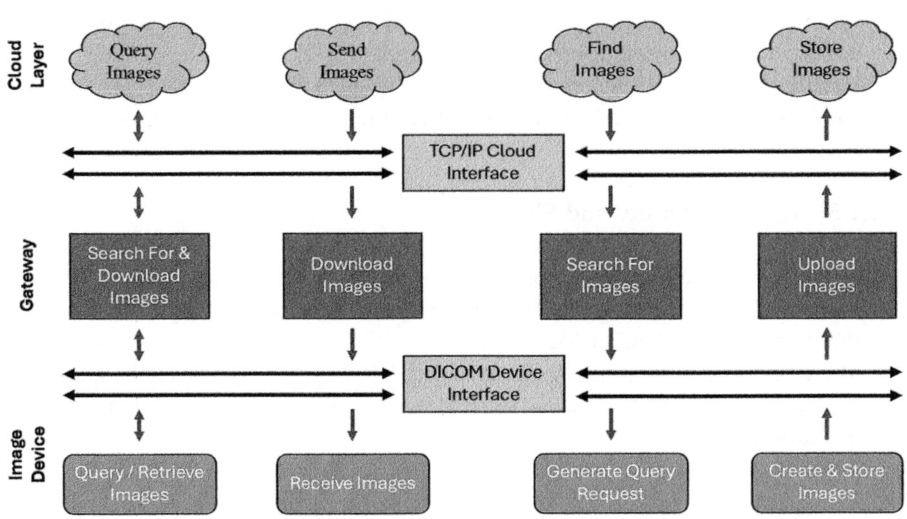


**Fig. 3** Imaging device to cloud service flow

### 12.2.1 Image Acquisition

- Medical imaging devices, such as CT scanners, MRI machines, X-ray machines, etc., produce DICOM (Digital Imaging and Communications in Medicine) images.
- These images are captured and stored in the hospital PACS (Picture Archiving and Communication System) or imaging modalities.

### 12.2.2 Medical Image Gateway Deployment

- The medical image gateway is installed on a server or virtual machine in the hospital network.
- The gateway is responsible for establishing a secure, encrypted communication channel between the on-premises PACS or DICOM-compatible devices and a cloud-based storage.

### 12.2.3 Data Transmission

- Once installed, the medical image gateway identifies new DICOM images in the local system.
- It compresses, anonymizes (if needed), and encrypts these images using DICOM standards and securely transmits them to the Cloud platform.

### 12.2.4 Cloud-Based Image Storage and Access

- Images uploaded to the cloud can be stored, viewed, and shared securely.
- The Cloud platform allows for on-demand access to medical images through web portals, API integrations, or mobile apps.
- Healthcare professionals can access these images remotely from any location, and patients can receive direct access to their medical images through secure links.

### 12.2.5 Interoperability and Sharing

- The medical image gateway ensures compatibility with other medical systems, such as EMR (Electronic Medical Records and Electronic Healthcare Records), PACS, and other cloud or local systems.
- The gateway also supports integration with EHR systems, allowing hospitals and healthcare providers to directly import/export patient records and images across institutions.

### 12.2.6 Complianceand Security

- Data is transmitted and stored with encryption, both in transit (using TLS) and at rest (using AES-256 encryption).
- The platform is compliant with HIPAA, GDPR, and other healthcare data security and privacy regulations, ensuring patient data is protected throughout its lifecycle.

### 12.2.7 Image Sharing and Collaboration

- Healthcare professionals can easily share images with colleagues or specialists for second opinions or consultations.
- The system supports image annotations and reports, making it easier to collaborate on diagnostics.
- The medical image gateway allows for web-based DICOM viewers, eliminating the need for specialized software on local machines.

### 12.2.8 Disaster Recovery and Scalability

- Since the data is in the cloud, the cloud platform provides automatic backups and ensures disaster recovery.
- The cloud-based architecture makes it scalable, so healthcare institutions can store an increasing number of images without worrying about local storage constraints.

### 12.2.9 Medical Imaging Gateway—Use Cases

- Telemedicine: Doctors can access patient imaging records remotely and collaborate in real time.
- Research Collaboration: Medical images can be shared across research institutions to facilitate large-scale studies.
- Patient Access: Patients can access their medical images securely via links, improving transparency and patient engagement.
- Cross-Institutional Sharing: Hospitals can share imaging data with specialists or different departments across multiple locations without requiring physical transfers.

## 12.3 Lack of Data Integrity & Accuracy

AI systems rely heavily on data for training and decision-making. If the data fed into an AI model is tampered with, it could lead to incorrect diagnoses or inappropriate treatment plans. This is especially dangerous in AI-driven triage systems, where patient priority may be skewed by inaccurate data.

**Risk Mitigation** Regular audits of data integrity are essential. AI models should undergo continuous validation to ensure accuracy. Furthermore, any data transmitted between cloud systems and local networks should be cryptographically signed to ensure that the information has not been altered. Cloud platforms provide such validation and audit mechanisms, ensuring that data remains accurate and trustworthy.

## 12.4 Cyberattacks on Connected Devices

Medical devices connected to the cloud, including AI systems used in radiology or robotic surgery, are susceptible to cyberattacks. An attacker could potentially gain control of a connected device, halting critical diagnostic or surgical processes, which could be life-threatening for patients.

**Risk Mitigation** Implementing robust network segmentation is a critical measure. By isolating AI diagnostic systems from other parts of the hospital network, any compromise can be contained. Additionally, cloud systems should be used to facilitate secure, remote diagnostics and updates, reducing the need for physical access to potentially vulnerable devices.

### 12.4.1 Compliance with GDPR & HIPAA Standards

Both GDPR and HIPAA enforce strict regulations concerning patient data privacy, storage, and access. Healthcare providers must ensure that their use of AI-based diagnostic tools adhere to these standards to avoid penalties.

**Risk Mitigation** To maintain compliance, healthcare organizations should use cloud services that offer audit trails, access logs, and encryption. Today, the cloud providers comply with both GDPR and HIPAA regulations, providing secure and compliant cloud infrastructure for storing and processing patient data. Additionally, healthcare providers should ensure they have clear policies around patient consent, data retention, and breach notification.

### *12.5 Remote Connections & Assistance*

Remote diagnostics and support have become increasingly critical in modern healthcare, especially in situations where onsite presence is limited. This is particularly true in AI-driven radiology and robotic surgeries. By enabling remote connectivity, healthcare professionals can troubleshoot systems, update AI models, and process diagnostic images without needing to be physically present. This improves operational efficiency, reduces response times, and ensures that healthcare services remain available even in emergencies.

**Remote Troubleshooting & AI Model Updates** Cloud-based systems enable secure remote access for troubleshooting AI systems, making it easier to fix issues quickly. In AI-assisted triage or robotic surgery, timely intervention can be the difference between success and failure. Remote connectivity ensures that critical updates to AI algorithms or software patches can be applied without delay, minimizing disruptions in patient care or even troubleshooting during operation.

**Risk Mitigation** Remote access should always be done over secure, encrypted connections, using multi-factor authentication and strict role-based access control (RBAC) including physical verification, such as YubiKey, face recognition in different facial expressions, voice recognition, and live location recognition as one of the verification methods, similar to what is being used in financial institutions. This ensures that only authorized personnel can interact with the system, reducing the risk of tampering or unauthorized access.

## 13 Conclusion

AI-driven diagnostic systems and robotic surgery devices offer transformative benefits in healthcare. By enhancing diagnostic accuracy, improving response times, and providing remote support, these technologies have the potential to revolutionize patient care. However, their integration into healthcare environments comes with significant cybersecurity risks. Healthcare providers need to address the identified risks by leveraging secure cloud infrastructures used by sub-tier suppliers who need to remain compliant with GDPR and HIPAA regulations as they implement robust encryption protocols and ensure remote diagnostic technologies remain updated. Lastly, as AI technology continues to evolve, playing a larger role in healthcare, cybersecurity must remain a top priority to protect both patient data and the integrity of medical procedures.

# Security Trends & the Evolving Digital Landscape

## 1 Medical Device Industry—Current Security Trends

What are security trends? Security trends are an ongoing battle between two entities, the proprietor and the external attacker, where the fight constantly escalates due to technological advancements, making it harder to know if you have been attacked. Specifically, the rough definition of security trends is evolving patterns, shifts, and emerging threats in the field of cybersecurity. So, when we examine this definition, how does it reflect on the medical device industry? How do we deal with the issues?

With a basic understanding of the security trends concept, what are the implications for the medical device industry, and how should the industry address the challenges? To start with, the medical device sector has adopted mainstream methods to combat the implications associated with security trends, ensuring that there is no loss of internet access, which largely resolves the issue. Further, adopting mainstream methods has resulted in a decrease in operational expenses, as well as a reduction in the stress associated with finding solutions to address cybersecurity threats. Those methods have been prevalent for years, but to remain competitive and to meet customer expectations, it is crucial to adjust to the modern era of cloud computing and data transfers. Consequently, it has become essential to connect medical devices to the internet, thus increasing the risks of cybersecurity threats.

With the necessity to have internet connectivity, medical device companies have embraced the approach of using local hidden wireless networks at hospitals. That point emphasizes that medical device companies delegate the management of servers to the hospitals. An example of that strategy is shown in the Vital Sync ™ system, which allows communications between local equipment and a nurse's tablet, used to monitor a patient's status.

When an update is needed for the hospital system, a medical device manufacturer sends a company representative with a flash drive to perform the updates. This method enables internal network communications between devices, which makes it

 
W. Harding et al., *Securing a Connected Future*,
https://doi.org/10.1007/978-3-032-07309-9_8

possible to maintain high security while incorporating updates to the existing systems without fear of outside security breaches. Furthermore, it is known that the proposed update method is not perfect because it still requires manually maintaining internal hardware. However, the method does reduce the chances of exposing a secure system to external threats.

But in a world where local servers are starting to disappear and the cloud computing era is rising, many cloud services follow HIPAA standards, providing a choice of where data will be stored and how it will be accessed. Specifically, sensitive data is secured using physical two-factor identification (Hilm and Rahim 2019).

## 2 Rapidly Evolving Digital Landscape & Its Impact on Cybersecurity

With the digital landscape evolving and new security trends emerging at a very fast pace, the medical device industry and the devices themselves become increasingly vulnerable to emerging cyber threats. Further, the integration of advanced technologies on both sides (i.e., target and attacker) makes it a constant race to have the most advanced defense or offense. For example, some of the key technologies that find their way into standard medical device manufacturing environments are Artificial Intelligence (AI), 3D printing or additive manufacturing, robotics/automation, Internet of Medical Things (IoMT), cloud computing, and Augmented and Virtual Reality (AR/VR).

From a threat perspective, even though today only narrow (weak) AI systems are available, the speed and precision systems are offered to attackers means that humans cannot respond to AI attacks, unless they possess the same technology. Accordingly, when we consider the future of AI and the eventual availability of general enhanced AI systems, the Alan Turing question from 1950, "Can machines think?" becomes a stark reality.

With the idea of "Thinking Machines" in mind, the following (non-exhaustive) list is an example of the potential impacts on both patients and the healthcare industry, which the reader should be aware of and potentially prepare to defend against:

- Functionality of medical devices.
- Delay patient care.
- Compromise sensitive patient information.
- Exploitation of vulnerabilities within the Internet of Things.
- Compromised patient safety.
- Financial losses.
- Reputation damage.
- Penalties from regulators.
- Disruption to operations.
- Increased mortality rates.
- Theft of intellectual property.
- Increased healthcare costs.

As we continue through this chapter, we will examine in more detail some of those listed impacts, but let's first look at some potential cyber threats that might lead to any one of those potential issues.

In recent years, one of the most common and devastating threats is a so-called ransomware attack. These attacks are a type of cyberattack where malicious software encrypts a victim's data, rendering it inaccessible until a ransom is paid to the attacker. These attacks are usually executed by well-organized crime groups or state-sanctioned attackers and often require a human to initiate the attack. Following the first occurrence of ransomware attacks, the threat level has increased substantially as the attackers have resorted to double or even triple extortion attacks. A double-extortion ransomware attack not only encrypts the data but also downloads it first, threatening to publicize anything that was acquired. And to make things worse, culprits might also go for any of the target's business partners or customers, which then makes it a triple extortion attack. That said, ransomware is part of a larger group of threats called malware (short for malicious software). Malware includes threats, such as viruses, worms, trojans, spyware, adware, and others. And though this book will not go into detail on all potential malware, it is important to consider these threats when planning any defense and response measures.

With malware and ransomware attacks in mind, the increasing interconnectivity of the healthcare industry and interoperability of medical devices represent a larger target, making them more susceptible to unauthorized access and corruption. For example, common incidents include the theft of intellectual property (IP) and sensitive data known as Cyber-Espionage. This can be achieved by exploiting vulnerabilities within an IoT ecosystem, such as across medical device manufacturing processes that use off-the-shelf software, through phishing attacks, or even through insider threats. Thus, existing cases of successful attacks underline our dependencies on suppliers and the dangers that come with the interconnectivity at a global level. And since attacks may start at third-party suppliers' locations, it is possible that those methods of attack are transferred to customers.

The vulnerabilities associated with third-party suppliers and globally connected technology represent a clear movement from closed-loop systems, where there are physical barriers to block intrusions and to an industry that is electronically separated from many forms of access. And with that point in mind, keeping data safe at the gateway, router, hub, and cloud storage level poses significant challenges. While these technologies offer numerous benefits, they also provide new vectors for cyberattacks. Unauthorized access to cloud-based systems can lead to data breaches and loss of control over medical devices and internal organizational systems. And since the healthcare industry has not embraced the use of standardized cybersecurity protocols, it becomes an increasingly difficult task to ensure consistent protection against cybersecurity threats.

Diving deeper into the impact that a successful attack represents to the user of a medical device, imagine the attack introduces malware that causes it to malfunction or deliver undesired therapy. An insulin pump, for example, could deliver an incorrect dose, pacemakers might fail to regulate a heartbeat, or a surgical robot could become uncontrollable during surgery. The patient's safety is at immediate risk if devices malfunction, with potentially devastating outcomes. Also, certain medical devices store and transmit sensitive patient data. And a cyber-attack can lead to the theft of that data, which can then be used for identity theft or sold on the black market. Furthermore, through previously mentioned device interoperability processes, devices transfer more of our personal information, such as health data, onto global systems and cloud storage, where a single data breach can expose the data of millions of patients!

New technologies and threats require rapid and effective responses, not just from the healthcare industry but also from governments and regulatory bodies. Specifically, governments and regulatory bodies need to support the healthcare industry to protect against potential threats and to develop new protection methods through the implementation of updated standards, regulations, and laws. For example, the FDA has developed guidelines recommending a lifecycle approach to cybersecurity for both pre–/post-market medical devices. In 2023, the FDA released the "Cybersecurity in Medical Devices: Quality System Considerations and Content of Premarket Submissions", outlining their recommendations for device design, labeling, and documentation to ensure devices are resilient to cybersecurity threats.

As for standards, one such international standard, ISO/IEC 27001, is widely recognized as a significant standard for information security management systems (ISMS). It is not a standard that is specific to medical devices but provides a comprehensive framework for the management of sensitive information, which includes health data. The standard aims to support manufacturers in understanding what it means to effectively protect data integrity, confidentiality, and availability. Further, the standard is effective in outlining the implementation of systematic risk management and security controls, where it is particularly relevant for connected medical devices that are integrated into global healthcare IT networks. Additionally, other standards reenforce and supplement the ISO 27001 standard, such as IEC 62304 (software lifecycle processes for medical device software), ISO 14971 (framework for risk management including identification and mitigation of cybersecurity threats), and IEEE/UL 2933-2024 TIPPSS (Trust, Identity, Privacy, Protection, Safety, and Security).

Mentioned previously, some technologies (e.g., AI and ML) can be used by both the target and attacker, where the following Fig. 1 explores some of the common applications for these technologies:

In general, AI/ML and automation are playing and will continue to play a significant role in the world of cybersecurity. Real-time monitoring and response are so important in today's digital world, which is dominated by automated intelligent systems. And because of the associated threats, any manufacturer of medical devices must consider the integration of security at all phases of device creation (e.g., R&D). Where consideration of cybersecurity threats must be embedded into the culture of all relevant stakeholders interacting with medical devices.

Within organizations a shift-left approach should be used, which means quality and testing strategies should move as close to the start of the development phase as possible. Including organizations keeping an accurate and detailed track of all the components within a device (e.g., vendor supplied, in-house development, updates,

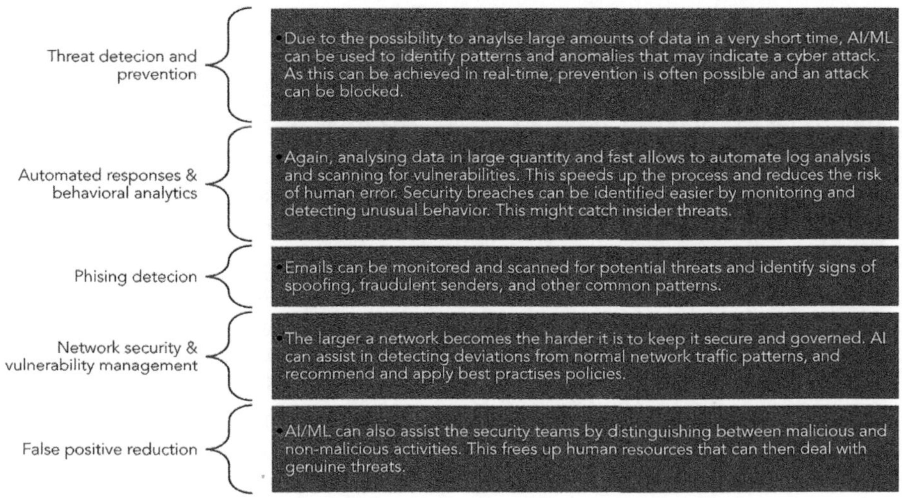


**Fig. 1** AI/ML cybersecurity applications

patches, etc.). It is crucial to have effective asset management which includes the use of Software Bill of Materials (SBoM), which captures all elements of integrated and deployed software. Organizations need to employ predictive analytic methods to enable them to anticipate potential cyber threats, scan for vulnerabilities, detect attacks, and respond to attacks. In other words, it is better to be proactive and prevent an attack than to expend resources reacting to and cleaning up after attacks.

Building on the point of expending resources, for organizations to enact preventative measures and to ensure robust product development processes, they must consider increasing security-related budgeting. The threat levels to medical technology are increasing substantially, and so must an organization's defensive actions. This is not possible using historical budgeting, which now requires careful consideration of resources such as specialists, as well as intelligent systems. Accordingly, existing investments should be optimized for the growing need of sophisticated attack prevention systems and resources channeled into specific cybersecurity initiatives.

## 3 Exploration of Use Cases and Potential Threats of Emerging Technologies

Expanding on the topic of ransomware attacks, the Conti Group is a well-known cyber-criminal organization specializing in ransomware attacks. Their attacks aim at infiltrating a target network, encrypting important files and databases, and then blackmailing the victim demanding a ransom to decrypt the files again. The Conti malware typically used during attacks is particularly effective since it is much faster in executing the encryption than most other ransomware.

### 3.1 HSE Ransomware Attack

In May 2021, using a Conti Group ransomware attack, cyber criminals successfully infiltrated the Irish Health System Executive (HSE) body. The HSE provides all the Irish public health services through hospitals and communities and is a central point of communication for all health care & service providers across the country. At the time of the ransomware attack, the HSE was in the middle of managing the COVID-19 crisis, and the attack could not have come at a more challenging time. After becoming aware of the attack, the HSE activated its incident response process, which eventually resulted in shutting down all its IT systems and disconnecting it from the National Healthcare Network (NHN). While this did cut off attacker access to the system, it also meant that all healthcare professionals in Ireland lost access to the HSE, including access to patient information, laboratory systems, clinical care, and others. Even systems, such as finance, payroll, and procurement, were rendered nonfunctional. The impact of the attack became obvious when healthcare professionals realized that the only way to still provide health care across an industrial nation with a population of over five million people was to revert to pen and paper. This meant that over half of all hospitals in the country had to cancel at least some of their outpatient appointments, if not all, and there were major disruptions to overall hospital operations. No emails, networked phones, or any other standard communication was available throughout all operational services, and to make things worse, major parts of the affected IT entities had been encrypted.

It might appear that attacks, such as the one HSE experienced have an immediate effect on the IT system landscape and that those attacks become apparent relatively quick. However, taking a closer look at the timeline and how the incident unfolded should raise awareness of the necessity for up-to-date security measures and rigorous, continuous monitoring (Fig. 2).

In the aftermath of the HSE incident, it became clear that besides the impact on operations, patient data and related information was also compromised where data

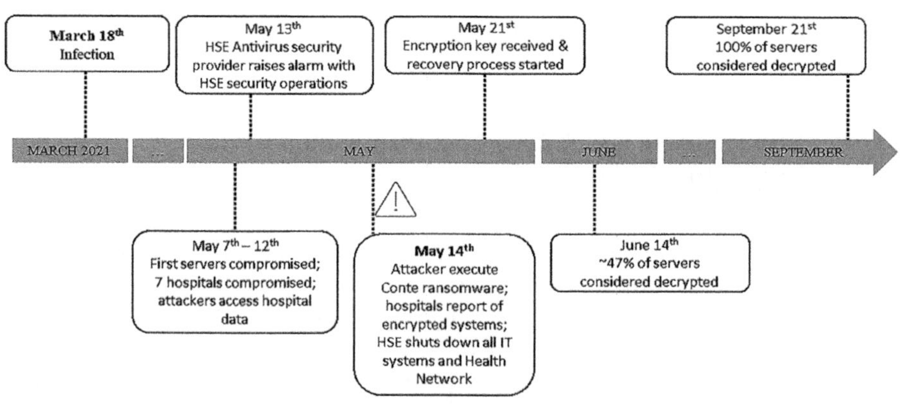


**Fig. 2** HSE ransomware attack—timeline (Hartmann 2024)

from ~113,000 patients were accessed and copied. The costs for this incident were estimated in the region of €53 Mio. for immediate responses and over €657 Mio. to implement improved cybersecurity systems. And though the attack was quite successful in the damage that it inflicted, it was a relatively simple assault. The techniques and tools used were well known and the time between infection and attack was used to fully exploit the system, resulting in the quick application of data encryption.

## 3.2 Potential Threats Posed by Security Trends

With the evolution of digital technology some of the emerging security trends create a whole new dimension of threats to the security of a company, its assets, and its ability to function. The arrival of generative AI is already putting huge pressure on the cybersecurity industry as it plans for what lies ahead. And when we pair the pressure with the immense computing power that quantum systems will deliver in the future, we might need to reinvent cybersecurity and think outside the box, where we will certainly see a new level of attack patterns.

Single or rogue attacks are rare today, where most incidents are coordinated campaigns with very clear goals. Strikes against IT systems are often organized by criminal organizations or governments and criminals have established a new modus operandi by providing tools and processes required to execute successful cyberattacks as a service. These organized criminal groups have discovered that they can sell their knowledge and tools with the help of standard business models, just like any other enterprise. One example is Ransomware as a Service (RaaS), which is a common platform and can be used by any antagonist (for a fee of course). RaaS is a cybercrime business model where ransomware developers sell or lease their ransomware software to affiliates, who then use it to carry out attacks. This model is comparable to legitimate Software as Service (SaaS) offerings, making it accessible even to those with limited technical skills. The business model even includes different revenue models, such as a monthly subscription, affiliate programs, one-time license fees, and profit sharing. And just as a legitimate SaaS company, RaaS providers often have a 24/7 support system, user reviews, customer forums, and different types of dashboards as well as tools for tracking infections, payments, and encrypted files. The extortion methods can range from a pure extortion model which exfiltrated data without encrypting it (threatening to publish it), double and multi-extortion, data encryption, and other attacks (e.g., DDoS), all executed to put pressure on the victims.

To combat RaaS attacks there are multiple measurements and best practices that you can take to protect your data and systems. These proactive measures include regular data backups, software updates & patches, employee training & awareness, advanced security solutions (e.g. endpoint protection), network segmentation, multi-factor authentication (MFA), and a general preparedness in case an attack occurs (e.g., preparing a detailed incident response plan).

## 4 Difficulties of Implementing Cybersecurity Measures—Case Studies

Wrapping up the discussion of security trends and evolving digital landscape as it applies to developing leading-edge medical device technology, we have suggested that cybersecurity should be a principal consideration that should be explored from design to deployment. Innovators need to be proactive with their inclusion of security measures, where it is suggested that innovators collaborate with both industry and academic experts as they define the best practices for enabling their designs to evolve along with potential threats. That said, we offer a few case studies that represent leading-edge technology, where weaknesses in applying cybersecurity measures have exposed vulnerabilities in the tech.

- Internet of Things (IoT): A 2016 Mirai botnet distributed denial of service (DDoS) attack compromised numerous IoT devices. The Mirai botnet DDoS attack also took down many websites associated with household consumer and entertainment/social media software. And it was also shown that IoT device vendors regularly neglect to apply appropriate security measures to their devices (Kolias et al. 2017).
- Autonomous Vehicles: A 2015 case where it was demonstrated that autonomous vehicles could be remotely controlled exposing a potential risk to vehicle occupants and others near the vehicle. Accordingly, autonomous vehicles represent compact standalone systems with an array of sensors, and as demonstrated in the 2015 case, future autonomous vehicles need anti-jamming technology, redundant sensors, and systems that are best described as misbehavior detection systems which can ensure that the vehicle does not detect false objects (Chowdhury et al. 2020).

## 5 Conclusion

As was explored and discussed in this chapter, there are many new technologies that can be used for the benefit of the medical device industry and ultimately patients. But with every emerging trend and new technology, there is the risk for misuse through exposed vulnerabilities or even brute force attacks. And for that reason, the reader was given a glimpse into how data and devices might become compromised and how those compromises put patients, healthcare professionals, and businesses at risk. Furthermore, through the examples and case studies presented, the reader has gained an improved perspective of how to recognize risks, to respond to threats, and how to protect themselves. That said, the reader was stimulated to establish best practices and to remain compliant with newly evolving regulations and international cybersecurity standards.

# The Future of Cybersecurity in Medical Devices

## 1 Visions of the Future

When thinking of what future cybersecurity might look like for medical device technology, there are some topics to examine. For example, we might consider (1) security measures, (2) vulnerability management, (3) threat monitoring, assessment, and mitigation, (4) data transformation and sharing, (5) IoT-enabled devices, and (5) AI/ML solutions. And those topics would show up in any list using a basic Google Scholar search. But what general searches do not expose is how the movements to using BYoD or even intra-body technology must be examined independently, where they should not be considered as covered by general IoT discussions. That point might appear obvious when we mention topics, such as data transformation, distributed systems, and IoT, but those are broad topics and do not emphasize that by embracing a BYoD or an intra-body model, solutions must be assessed through a different lens. For example, the creation of an open-air IoT solution presents a unique set of circumstances that a Biological IoT (BIoT) solution might need to consider. Further, the use of a BYoD model could manifest during any discussion of distributed systems as well as within the scope of AI-enabled solutions. But that discussion would not be forward-thinking enough, whereas the methods for transferring, aggregating, transforming, and even data storage is not the same as is presently used for most medical device technology (Soubhagyalakshmi and Reddy 2023).

Intra-body devices are also emerging as unique methods for establishing connectivity, transferring data, storing data, and even powering associated intra-body technology (Giftakis et al. 2025). Some examples of forward-thinking intra-body technology might include fractal antennas, the use of neuro pathways to transfer/collect data, sub-dermal transceivers, and energy harvesting. With those examples in mind, methods of data transference within intra-body solutions would not be the same as we might find with traditional open-air IoT solutions. Plus, though this book's topic is focused on cybersecurity for medical devices, it is important to



W. Harding et al., *Securing a Connected Future*,
https://doi.org/10.1007/978-3-032-07309-9_9

consider that any development of an intra-body solution might present more issues than those associated with just cybersecurity. Furthermore, the security of a patient/user may represent a higher risk than the risks exposed in a cybersecurity evaluation. That said, just the transference of data within an intra-body solution might manifest thermal issues that could damage patient tissue. And that issue is just scratching the surface of the issues/barriers that must be overcome when developing a safe intra-body solution (Vizziello et al. 2023). And, when we combine the concepts of intra-body with BYoD, we create an architecture that is dramatically different from what has been presently implemented.

Whether the discussion is on BYoD or traditional medical devices, emerging trends and technologies in medical device products are challenging current cybersecurity concepts. A big trend is the use of AI, and when integrating that into medical devices it can be very useful in helping to detect a heart attack earlier or even prevent a mis-shock of an implanted defibrillator. On the flip side, there can also be vulnerabilities within the algorithms which can be exploited if not secured properly. Even wearable devices that track your heart rate and other sensitive health data necessitate robust encryption and data protection measures.

Another emerging trend is the use of telemedicine and remote monitoring. The use of remote healthcare solutions is on the rise as a method for reaching patients in remote areas, as well as a method for transferring data more quickly between patients and healthcare providers. However, when using remote monitoring and telemedicine technology, there is a need to use secure data connections. And though telemedicine is mostly focused on collecting and transferring patient data to and from healthcare institutions, there is a need to also consider data that runs the telemedicine technology as well as the data that institutions use for billing, management, IT support, quality control, etc. For example, healthcare institutions cannot run without a functional financial, facilities, or IT department, and if any one of those critical departments were to fail, then the ability of the institution to support patients would be at risk.

As we establish a vision of the future for cybersecurity in medical devices, some use cases come to mind, specifically as we think about wearable tech or tech that is integrated into environments that have not been previously considered. For example, consider military uses, technology used by astronauts, or even the integration of a human with vehicles and homes. With those ideas in mind, the future of medical device cybersecurity could be better understood, by looking at that future through the eyes of Johnny Mnemonic (a cybernetic data courier from the similarly named 1995 movie). The inclusion of cybernetics, robotics, intelligent prosthetics, wearables, implantables, and biocybernetics is also in our near future (Harding et al. 2021). The future of medical device cybersecurity must move from a simple examination of hardware and software items, to include the human and more specifically biological elements (e.g., human, animal, plant, etc.). The next sections will promote awareness of evolving technology trends and medical devices/systems that need to be evaluated through the lens of cybersecurity that protects both the human and the ideas of humans.

## 2 Emerging Trends & Technologies in Medical Devices—Cybersecurity Concepts

As previously discussed in chapter "Security Trends & the Evolving Digital Landscape" regarding the Irish Health System Executive ransomware attack, there was no evidence of any compromised, networked medical devices, but the bi-directional flow of information was not possible due to the National Health Network shutdown (Daly 2022). The implications of such an event are yet to be fully understood, but it does not take a lengthy analysis to comprehend the lethal potency of such an attack. The inability to communicate with interconnected devices because of a complete system shutdown was merely a by-product of an intrusion in the wider ecosystem.

Building further on the topic of interconnectedness, and as suggested by Lewy (2015) there is a need to make healthcare and therapies more accessible, less intrusive to the body, and easily replaceable, while promoting the creation of a 360-degree picture of the patient (e.g., wearable technologies). For example, wearable devices are no longer self-contained or closed loop systems, but instead they promote bidirectional data exchanges, such as to/from the device and the Health Care Providers (HCP). However, interconnected systems hold risks, such as the introduction of new vulnerabilities (Abdullah and Jones 2019), tampering or unauthorized access, data theft, systems encryption, external device interference (Pycroft and Aziz 2018), manipulation of virtual environments, quality impacts and product counterfeiting.

For example, the following list outlines some of the rising technologies and therapies that will challenge our current cybersecurity standards and impact our understanding of Data Protection and Privacy (DPP):

- Interconnected medical devices and systems (Medical Internet of Things—MIoT).
- Remote patient monitoring.
- Integration of Artificial intelligence and Machine learning.
- Human-computer interfaces (brain chips).
- Complex implantable medical devices.
- Closed-loop medical devices.
- Telemedicine and connected healthcare ecosystems.
- Augmented Reality (AR) and Virtual Reality (VR) in healthcare.
- Robotic surgical platforms.
- 3D Printing of Medical Devices
- Blockchain in Healthcare.
- Biometric Authentication.

# 3 Future Systems & Technology Requiring Robust Cybersecurity

Moving from our discussion of emerging cybersecurity trends, intra-body communications, and interconnectedness, we examine some less obvious and more forward thinking healthcare technologies that require the integration of strong cybersecurity.

## *3.1 Virtual Healthcare Assistants*

Though the concept of an AI-enabled virtual healthcare assistant (VHA) might spark images of a robotic doctor or nurse giving a human patient a shot or taking their blood pressure, the true image of an AI VHA might be nothing more than an input device like Amazon Echo or Google Home. Of course, those 2 input devices represent just the top level of an AI VHA, which can work interoperably with IoT devices, hospital equipment, healthcare professionals, patients, emergency response resources, EHR/EMR data, and cloud-based AI systems. AI-enabled VHA requires minimal human intervention and exists to promote faster and more accurate individualized health diagnostics, monitoring, therapy, and improved health outcomes (Fig. 1).

Considering that an AI VHA is an AI-enabled tool that possesses specific patient data as well as access to patient-centric technology, care must be exercised to ensure that patient data is protected without compromising the quality of care provided to the patient. Additionally, it is important to keep in mind that an AI-enabled VHA might also be represented by a system that assists a surgeon in performing a process

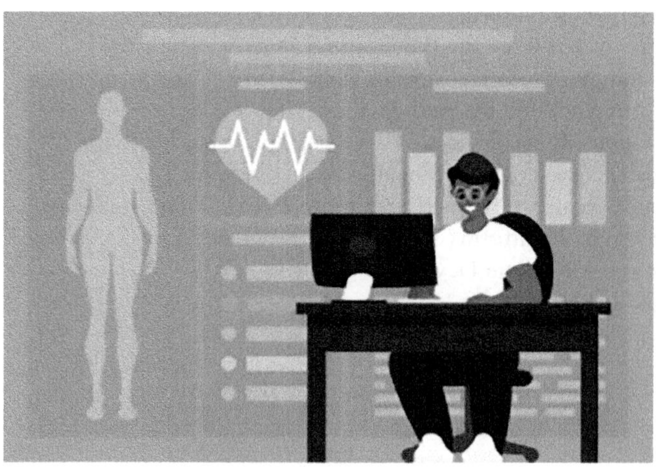

**Fig. 1** Virtual health assistant (Free-Vectors.Net)

or helps hospital resources manage patient claims. Consequently, the first AI-enabled VHA might be best represented by AI ChatBots that have access to vast amounts of health-related data and can act as the initial interface for humans seeking nonlife-threatening healthcare assistance.

With the discussed types of AI-enabled VHAs in mind, it should be obvious that such a connected and personalized device needs to have its physical and data elements protected from intentional as well as unintentional access, misuse, and disruption. Accordingly, the primary components of an AI-enabled VHA that need to be protected are patient data, system integrity, and system availability. And as far as the protection measures of an AI VHA that should be examined are access, threat detection, threat response, data storage, and data transmission. Lastly, and less obvious aspects of an AI VHA that need to be considered are misinformation and bias as they relate to compromised data that could negatively impact a patient's quality of care.

## 3.2 *Neuromorphic Computing*

Neuromorphic Computing is just a scientific term for digital designs that mimic a human's neural networks. And though the most current perspective of neuromorphic computing is on creating machines that simulate human brain functions, the original intent was to create digital components that could be integrated as an intrabody mesh that promoted communications and power distribution within a human body (Giftakis et al. 2025). Furthermore, the concept of neuropathic computing is to collect data, transform data, and use digital data to improve a human's quality of life. Thus, it becomes clear that if data is being moved intrabody, then we must consider all potential risks associated with disruption of that data flow. Where it is proposed that intrabody data must be encrypted, such that accidental and intentional disruption can be detected and resolved (Fig. 2).

With the previous points in mind, it might not be clear what neuromorphic computing is and how it is represented. So, we created a device patent (Harding and Maluf-Burgman 2021), which is a system of fractal antennas, conductive contacts, and software that enables implanted devices to transmit signals within a biological environment. More specifically, our neuromorphic solution uses a patient's nervous system as a communication conduit. The solution uses specific neurological pathways to move data from one point to another through the brain stem, the reticular formation (i.e., a region in the brain stem consisting of hundreds of small neural networks), and then on to the thalamus, eventually ending up in the appropriate sensory area located in the cerebral cortex. And since the patented device receives instructions and inductively coupled power, it is easy to see that without implementing cybersecurity procedures and methodologies, there is a high risk to a patient's quality of life.

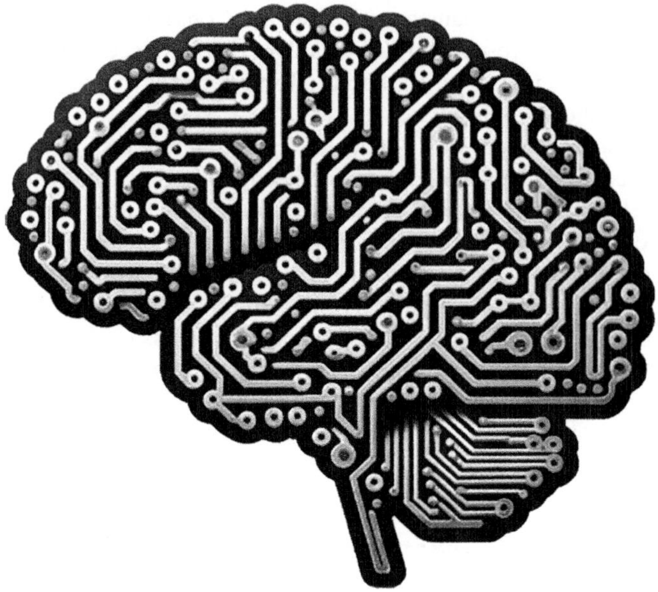

**Fig. 2** Neuromorphic computing (Wikimedia Commons.Org)

## *3.3 Spatial Computing*

When thinking of spatial computing it is natural to quickly envision the space around us and how we might use computing devices to understand our environment. And if that is truly the first thought that pops in your mind, we can tell you that you are not too far off the mark. Further, it might surprise you to learn that you have most likely already used spatial computing. For example, if you have ever used augmented or virtual reality tech, then you have certainly used spatial computing technologies (e.g., Pokemon Go).

If the term spatial computing does not really trigger any deep thoughts about the technology embodied by the term, we offer that the term eXtended Reality (XR) is the best way to represent the concept of spatial computing. And since it represents augmented (AR), virtual (VR), and mixed reality (MR) technology, you could imagine wearing a headset that enables you to see around you, but where your vision is enhanced by sensor information. For example, imagine configuring a surgical setting/procedure using XR-enabled technology (see Fig. 3) and you see the other human's vitals (e.g., heart rate, blood pressure, temperature, etc.).

Building on what you have learned about spatial computing, imagine that the sensing data that you are seeing while looking at the person were to become compromised, disrupted, or even hacked. And then imagine that you are a doctor examining a patient while wearing your XR headset. What happens if you make the wrong diagnosis based on hacked data or if you dictate a prescription that is filled

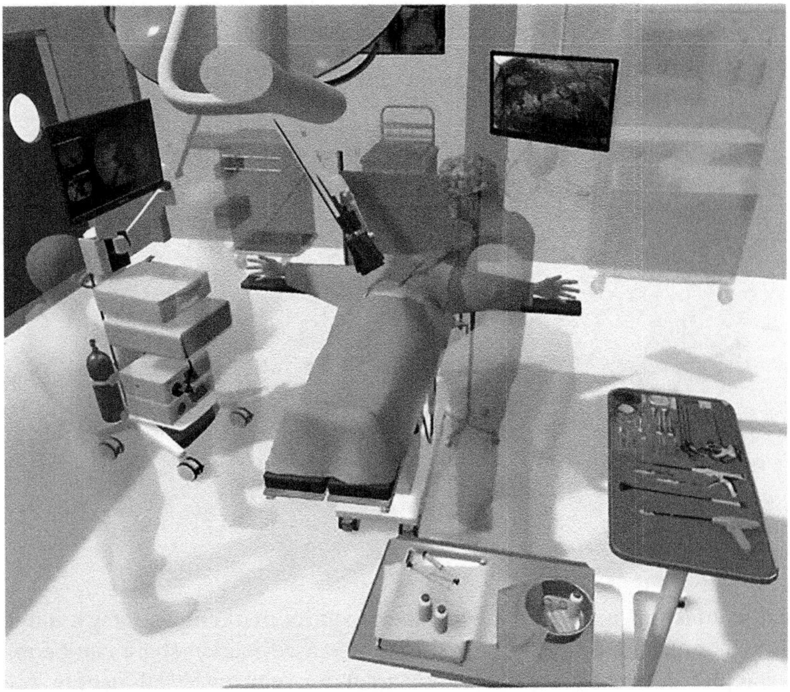

**Fig. 3** Spatial computing—surgical suite (Harding 2021)

incorrectly, because the data was altered? Patient poor health or death are potential outcomes if data gathered or transmitted is compromised.

All things considered spatial computing technology offers medical device designers with a platform that enhances innovation and communications, improves virtual meetings with healthcare professionals, and promotes methods for quick patient diagnostics within a virtual point of care environment (Gertenback et al. 2022).

Thus, the need to ensure that spatial computing data is shielded from interferences and secured against hacking becomes obvious when weighing the risks to patients. For that reason, XR tech needs to be designed, assembled, deployed, and used with cybersecurity as a principal thought.

## 3.4 Internet of Behavior (IoB)

Building on the connectivity aspects of IoT devices, an Internet of Behavior (IoB) solution seeks to leverage mixed technology, data sets, algorithms, and behavioral science to create a holistic human profile. With the goal of gathering multifaceted insights from disparate data collection instruments, the resulting IoB profile(s) can

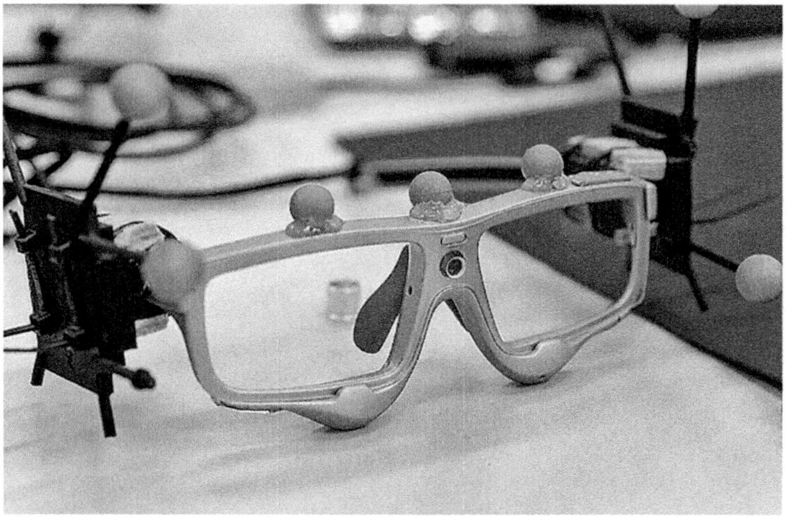

**Fig. 4** Internet of behavior (Openverse.Org)

be used to construct a medical device manufacturer marketing strategy and to optimize the manufacturer's budget, such that increased market share can be realized. For example, data that represents eye tracking, physiological pattern tracking, touch/click/swipe events, study demographics, etc. can be gathered from AI/ML vision sensing, AR/VR interaction, and the use of mobile devices. Thus, resulting in a vast amount of phenomenological and behavioral information that represents endless marketing insights (Fig. 4).

Though the tools used to create an IoB ecosystem are not unique, the methods of combining the tools will be unique to the initiating organization. Therefore, it will be necessary to protect data associated with designing a unique healthcare research study, participant/patient identities, analysis methods, marketing strategies, financial information, and resulting holistic user/patient profiles. All aspects associated with creating the healthcare research study and results should be treated as trade secrets, given that the resulting data will be used by the initiating organization to gain market share advantages as well as a better understanding of the patient and healthcare professional's needs.

Of course, if you have ever wondered why, you are emailed or texted advertisements for things that you might have searched through Amazon or Google searches, it might be clear to you that anything that you do on the internet and even things that are looked at on social media sites is being tracked. Now consider a tool that could construct a very accurate profile of you and how that profile could be used for good as well as bad outcomes. For example, if you were looking up a particular vitamin or home healthcare device, you might be pointed to a great price on the exact product that you need. Conversely, imagine that an accurate profile of you has been constructed, such that a system can predict when you would not be home or predict the password to your banking account.

With those points in mind and considering the power represented by an IoB solution, it is very clear that the data used, and the resulting profiles created must be protected at the highest level and that a robust cybersecurity solution should be put in place. Not only to protect humans but also to protect any medical device manufacturer who invests large sums of money to develop IoB solutions. Where it is obvious that the human should be protected but it is equally obvious that the associated intellectual property (IP) must also be protected from infringement as well as misuse.

## *3.5 Emotional AI*

The concept of AI does not immediately evoke thoughts of high energy, emotional, or passionate expression. And why should it? AI is something that we equate to a process or technology that intelligently responds to questions, interprets visual data and offers analysis, or helps us perform tasks more quickly and with less human error. Specifically, we do not associate AI with emotions, and that is the nature of our relationship to AI. But what if AI could be applied in such a way as to enable technologies to respond to emotions, to recognize variations in vocal tones, and to react to other contextual cues that humans exhibit? Would such a technology be helpful in the healthcare industry, and would it present risks that might be undesired? We believe that the answer is 'yes' to both parts of that question.

For example, as a person who has had exactly 34 surgeries, mostly because of engagement in high-risk activities, it was often the case that healthcare professionals had a difficult time assessing my pain threshold. And since my tolerance to pain is higher than normal and I metabolize drugs faster, administering pain therapy was a difficult task for healthcare professionals. But what if there were a system that could create a baseline of my uniqueness, recognize both verbal and nonverbal cues, and assess my needs more objectively? Could such a technology have helped me reduce negative events that often followed a human's biased assessment of my situation (i.e., based on an assessment of other patients)? I may never know, but I would venture a guess that it would have helped.

With that information in mind, we propose that through the lens of AI enabled technologies, that AI technology (i.e., vision and acoustic sensing) will exist that will use the vast amount of data contained in LLMs and data acquired through direct evaluation of a patent, to create an emotional AI. A technology that is perceptive of our opinions, our views, our tones, our facial expressions, and our gestures. Or more specifically, an effective computing solution that combines machine learning (ML) with AI that is more integrated with and invested in our psychological well-being. In a sense, the emotional AI technology would be an empathetic tool much like a human friend (Fig. 5).

And like a human friend, an emotional AI solution would inherently establish a trust relationship with us and would know things about us that we might not want others to know. Further, such a system could be used to do harm if misused or

**Fig. 5** Emotional AI (Openclipart.Org)

hacked, where the damage to a human's emotional state might have deeper and longer lasting negative consequences. And when considering the level of damage that an emotional AI could do to a human, it makes one wonder if it is worth the risks. We propose it is worth the risks since emotional AI enabled technology would provide us with healthcare solutions that move beyond the concept of one-size-fits all. But, like any technology that has access to patient or other protected data, we must build emotional AI systems starting at the ideation phase, always aware of the risks associated with compromised data and always with the need for data protection in mind.

### 3.6 *Implantable Microchips*

The idea of implanting a microchip sounds odd since it would be a foreign object, but it is more common than you might think. And when you think about how many microchips humans are already wearing, the move to implanting surface level devices might be the next evolutionary step for humans. For example, many people have implanted pacemakers, pain therapy devices, drug pumps, health monitoring devices, and devices that stimulate regions of the brain (DBS). And some of those devices can talk to externally worn devices or monitoring systems. Similarly, people with diabetes often wear a sensor with a small needle that breaks the skin and monitors glucose levels, transmitting that data to wearable insulin delivery pumps that also have needles that enter the body.

So, when we think of what future implanted microchips might represent, they could be bioinert technology that is able to power itself using energy harvesting methods or through passive inductive coupling. And in the case of inductive coupling (like you do when you place your iPhone on a charging pad), the implanted devices could use the ambient energy that surrounds us to continuously charge themselves or to receive enough power to perform a single diagnostic action. Furthermore, if you had a small passive microchip under your skin (like the RFID tags that we put in dogs and cats), you could power that microchip through your Apple Watch, which might be placed over the area where the microchip is implanted. Then, when you wanted to know your glucose level or test for other blood borne elements, you would only need to tap your watch or mobile device and it would power the implanted microchip and gather the data that you want (Fig. 6).

As an example of what an implanted microchip might look like (see Fig. 7) and how it might function, we created in 2013 the Injectable Sensor System for Measuring and Communicating Biological Data (or ISSM-CBD). The sensor device provides a method for passively collecting data for extended periods from within a patient's vascular system without the need for invasive monitoring/sensing methods, such as are normally associated with needle sticks and costly external measuring/sensing equipment (e.g., for detecting blood glucose levels). And by providing a biocompatible device using stent deployment methods the device acts as a platform for multiple sensors, such as would be used to monitor blood flow/velocity, temperature, oxygen levels, ejection fraction, glucose, etc. the ISSM-CBD. Furthermore, the device can be injected into a patient (i.e., during an outpatient visit, just under the skin) and passively charged (inductively) using RF energy to create a bidirectional link where an external monitoring/wearable device (e.g.,

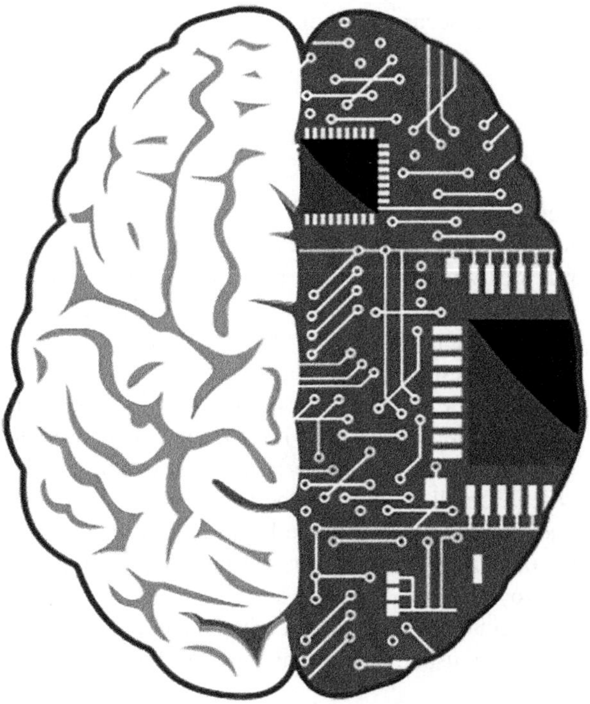

**Fig. 6** Implant microchips (Openclipart.Org)

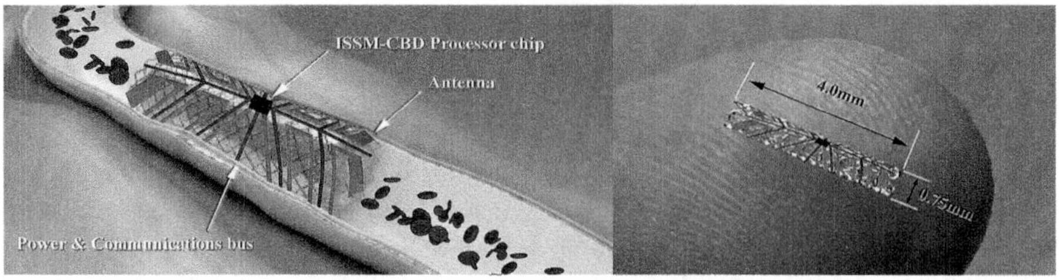


**Fig. 7** Injectable sensor system for measuring & communicating biological data (Harding 2018)

NFC enabled diabetes pumps) can communicate, control, and collect biological data from single or multiple ISSM-CBD devices.

Considering that the above discussed ISSM-CBD would receive power inductively and that it would transmit data that would be used to make health related decisions, we know that the device would need to have embedded and upgradable cybersecurity software and protective circuits. For example, imagine that the device was to receive a higher amount of power than is needed, resulting in a possible risk to the patient related to an overcharge. Additionally, imagine that the device was communicating with a continuous glucose monitoring (CGM) and delivery system, it would be easy to see how corrupted data might result in the patient not getting the required dosage of insulin or the patient might receive a higher than needed dosage. For those reasons (and a few more) the proper integration and the interoperability of the ISSM-CBD with other systems and devices would require an appropriate amount of security and protection.

## 3.7 Genetic Engineering

For at least the last 5 years, viruses have been at the top of the list of discussed topics. And as scientists work to limit viral events, the ability to strengthen the immunity of living things (e.g., human and non-human) is a critical need. Accordingly, the science of genome editing (also called gene editing and genetic engineering) gains momentum, specifically as a process that scientists use to change DNA. The genetic engineering process allows genetic material to be inserted, modified, or deleted. One such process is called CRISPR-Cas9 (clustered regularly interspaced short palindromic repeats) associated with protein 9. CRISPR-Cas9 mimics the genome editing that bacteria use. For example, when infected by a virus, bacteria capture small amounts of the DNA from the virus and insert those portions into the bacteria's DNA so that the bacteria, in effect, remembers the virus. If the bacteria encounter the virus again, the bacteria use that stored information to attack and destroy the virus (Fig. 8).

Of course, genetic engineering opens another can of worms as it relates to creating, analyzing, using, and storing data. And as with any process that uses data, there is a

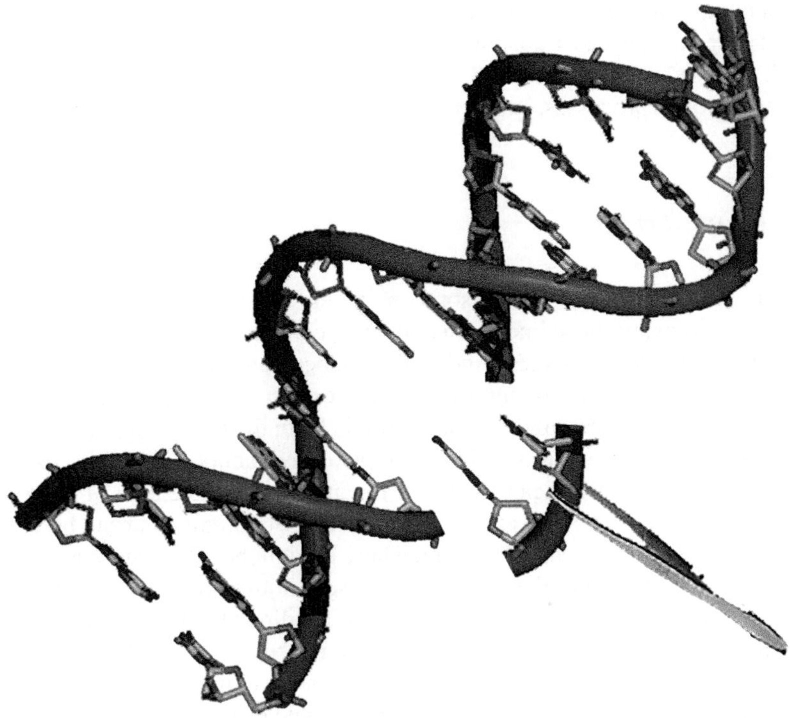

**Fig. 8** Genetic engineering (Wikimedia Commons)

need to ensure that the data is protected and not misused. For example, misuse or compromised genetic engineering data could lead to the creation of bioweapons. With that point made, it might not be clear how genetic engineering data is created, used, or stored. Accordingly, the following examples are provided to help increase awareness of the risks and the need to protect data found within the field of genetic engineering.

- Data Creation: DNA sequencing (e.g., Sanger sequencing) is the action that is first performed to enable scientists to identify the order of nucleotides (i.e., DNA building blocks A, T, C, and G) and protein sequences. Additionally, the methods, processes, and formulas used to create genomic data represent intellectual data that needs to be protected.
- Data Usage: Genetic engineering data is used to promote an understanding of gene functions as well as variations. And from that understanding, scientists can use genetic data to construct diagnostic tests for diseases, to identify individuals, such as for solving crimes, and the genetic data can be used to store data in DNA sequences.
- Data Storage: Besides the ability to store data in genetic material, the need to securely store genetic engineering data is a high priority, specifically because of the potential to weaponize and misuse the data. And with there being nearly 3 billion base pairs, the human genome requires substantial memory to store the genetic data.

In summary, in the field of genetic engineering, there is a need to provide secure methods for storing and transmitting genomic data, such that it cannot be intercepted, misused, or compromised. And just as the data needs to be protected, the ideas, techniques/processes, as well as any created organisms, require protection.

## *3.8 Drone Triage*

It is very common to see on TV someone using a drone to capture images of an event and to transmit those images in real-time, enabling the user to make decisions based on the collected images. And with advancements in military systems (e.g., unmanned) we see drones being used as devices that deliver devastating weapons. But a potentially new area of drone usage might be in the monitoring, diagnosis, and delivery of healthcare solutions. For example, why cannot a drone be used to fly medicine to a native American living off the grid in a remote area of Arizona? Or even better, could not a semi-autonomous or fully autonomous drone be used as a triage platform in a mass casualty event, where it might be too dangerous for another human to enter, but where the drone could assess injured individuals through non-contact sensing (e.g., UV, IR, Thermal, RGB, directed acoustics, etc.). Then, depending on whether survivable individuals have been identified, the drone could be used to communicate with the injured and possibly deliver some level of medical assistance (e.g., like firing a dart with an appropriate drug).

Considering the use of drones to triage injured individuals and deliver healthcare related therapies, several risks need to be considered. Some of those risks might be malicious attempts to interfere with the function of the drone (e.g., signal jamming, destruction, or system corruption), which would inhibit the drone from performing desired functions. And an even more dangerous situation might be the hacking of a drone, such that the drone is instructed to deliver lethal drugs. For example, consider the scenarios where a drone has performed triage on a hiker in a hard-to-reach area, and determined that the hiker is experiencing anaphylactic shock, but instead of delivering adrenaline, the drone delivers a lethal amount of morphine. Or consider that a drone has been sent to assess an IED event, and instead of correctly determining that some of the injured can be saved, the drone determines that no one survived the IED attack. How do we keep those negative events from happening? (Fig. 9)

Any technology that will be used to provide even the slightest amount of healthcare must be built from the foundational level up, with safety, security, and protection in mind. When a triage drone is designed, developed, tested, deployed, and sustained, that drone must have every aspect of its structure and function examined for potential risks. Technology like triage drones must be built with the assumption that corruption or misuse will occur and that corruption might not always be intentional. Builders of triage drones must create robust cybersecurity solutions that can detect, monitor, and respond to any attack (i.e., intentional or unintentional).

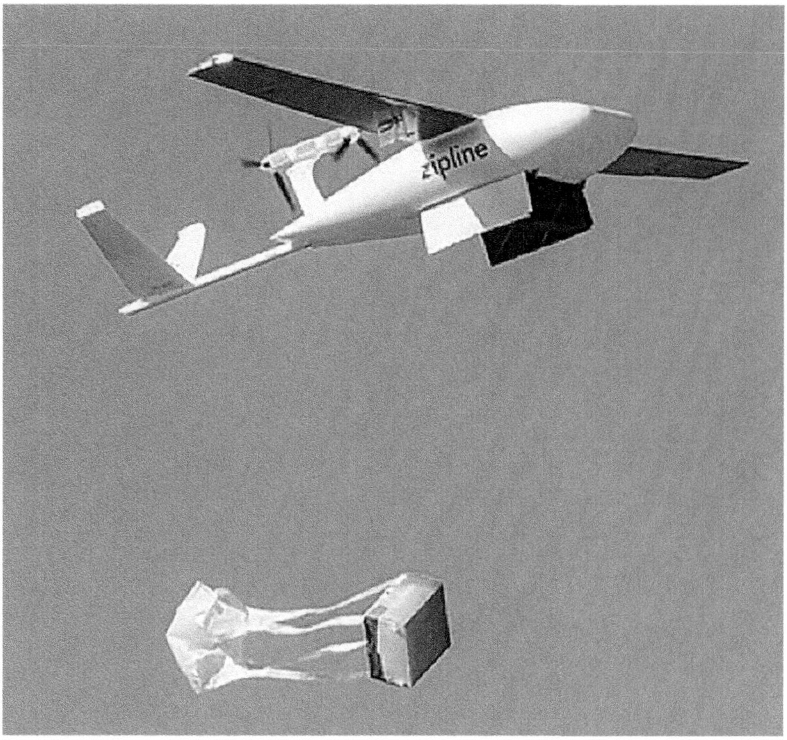


**Fig. 9** Drone Triage (Wikimedia Commons)

# 4 Conclusion

An examination of the technologies that were presented in the Future Technologies and Emerging Trends section might stimulate thoughts that the technology is far into the future, but what might be surprising to learn is that many of the technologies already exist. The technologies are still evolving, and a few have not reached widespread acceptance. What surprises us the most is that many of those futuristic technologies have not been properly integrated with cybersecurity measures. This emphasizes the fact that many emerging technologies do not consider cybersecurity until after an attack or at the point of technology deployment.

Additionally, the issue with implementing a cybersecurity methodology to protect data, the user, and your intellectual property (IP) is that it requires planning and must involve relevant stakeholders during all phases of technology ideation, decision-making, design, development, deployment, testing, sustaining, and decommissioning of the technology. More specifically, without engaging experts in the cybersecurity field and involving the end users/customers in the process's technology integration, any new technology will likely fail to protect the data that it consumes, transforms, and transmits. And if technology fails to protect data and IP, the ramifications are loss of market share, loss of customer confidence, and reduced quality of care that a healthcare customer needs.

# Final Thoughts

The future of Cybersecurity and its impact on the medical device industry is clear. Cybersecurity threats will only continue to grow as they are a great source of funds for attackers. According to Reuters, medical data is 10 times more valuable than credit card information on the black market (Wagstaff 2015), and as was presented in this book, the Reuters proposal has been substantiated. Cybersecurity threats are not going away, and medical companies as well as healthcare organizations (i.e., users of medical technology) need to heed the warnings by investing more time and money in creating robust future-forward cybersecurity infrastructures, where detection, identification, and response are the primary objectives. This suggestion not only includes guarding against direct attacks, but also considers attacks that are indirect and may originate via third-party collaborators/partners or from within an organization.

Despite increasing complexity in cybersecurity threats, the future of medical technology is exciting, especially when examining the future through the lens of IoT and BYoD technology as well as AI-enhanced solutions. With the integration of IoT solutions into the medical device industry, patients and organizational IP are at greater risk due to the vulnerability of their data. As the medical device industry starts to rely on the interoperability of IoT-enabled solutions, greater attention must be paid to weighing the benefits of using advanced technological solutions against the risks to a patient's quality of care because of increased cybersecurity attacks. A mindset that must be embraced is that the medical industry needs to move away from working in silos and start to learn from partners and even competitors, resulting in improved abilities to defend against cybersecurity threats by working together. There are cooperative methods of sharing data, such as those found in federated learning models, that promote sharing information while still protecting an organization's data, its patients, and ideas. Similarly, medical device companies need to invest more time and money in training employees, specifically on the topic of cybersecurity risks. The medical device industry needs to have cybersecurity at the forefront of its design, deployment, and sustainability efforts, with a clear focus on mitigating cybersecurity attacks, long before the attacks occur.



W. Harding et al., *Securing a Connected Future*,
https://doi.org/10.1007/978-3-032-07309-9

# References


Abdullah T, Jones A (2019) eHealth: challenges far integrating blockchain within healthcare. In: 2019 IEEE 12th international conference on global security, safety and sustainability (ICGS3). IEEE, pp 1–9

Adesokan-Imran TO, Popoola AD, Ejiofor VO, Salako AO, Onyenaucheya OS (2025) Predictive cybersecurity risk modeling in healthcare by leveraging AI and machine learning for proactive threat detection. J Eng Res Rep 27(4):144–165

Barrera D, Bellman C, Van Oorschot P (2023) Security best practices: a critical analysis using IoT as a case study. ACM Trans Privacy Secur 26(2):1–30

Blanton S (2024, October 16) Recent Ransomware attacks in 2024 (Updated Nov 2024) jumpcloud. https://jumpcloud.com/blog/ransomware-attacks-in-2024

Buermann G, Perucica N (2023, May 19) How can we combat the worrying rise in the use of deepfakes in cybercrime? World Economic Forum. https://www.weforum.org/stories/2023/05/how-can-we-combat-the-worrying-rise-in-deepfake-content/

Carreras Guzman NH, Wied M, Kozine I, Lundteigen MA (2020) Conceptualizing the key features of cyber-physical systems in a multi-layered representation for safety and security analysis. Syst Eng 23(2):189–210

Carroll M (2025, June 25) Patient's death linked to cyber-attack on NHS, hospital trust says. Sky News. https://news.sky.com/story/patient-death-linked-to-cyber-attack-on-nhs-hospital-trust-says-13388485

Cheng L, Liu F, Yao D (2017) Enterprise data breach: causes, challenges, prevention, and future directions. Wiley Interdiscip Rev Data Min Knowl Discov 7(5):e1211

Chowdhury M, Islam M, Khan Z (2020) Security of connected and automated vehicles. arXiv preprint arXiv:2012.13464

Chui M, Malhotra S (2018) Notes from the AI frontier: AI adoption advances, but foundational barriers remain. McKinsey & Company. https://www.mckinsey.com/~/media/McKinsey/Featured%20Insights/Artificial%20Intelligence/AI%20adoption%20advances%20but%20foundational%20barriers%20remain/Notes-from-the-AI-frontier-AI-adoption-advances-but-foundational-barriers-remain.ashx

Cola C (2020) Is the Coca-Cola formula kept secret because the company has something to hide

Cybersecurity & Infrastructure Security Agency (2022, February 10) Ransomware awareness for holidays and weekends. Cybersecurity & Infrastructure Security Agency. https://www.cisa.gov/news-events/cybersecurity-advisories/aa21-243a




W. Harding et al., *Securing a Connected Future*,
https://doi.org/10.1007/978-3-032-07309-9

Cybersecurity & Infrastructure Security Agency (2023, June 6) About CISA. Cybersecurity & Infrastructure Security Agency https://www.cisa.gov/about#:~:text=the%20American%20public.-,CISA's%20Work,infrastructure%20security%2C%20and%20emergency%20communications

Daly P (2022) "Writing on a curved surface" the operational response to the cyber-attack on the Irish health service. Médecine de Catastrophe-Urgences Collectives 6(4):275–277

Darktrace (2021, December 2) Darktrace reports 30% more ransomware attacks targeting organizations during the holiday period. Darktrace https://darktrace.com/news/darktrace-reports-30-more-ransomware-attacks-targeting-organizations-during-the-holiday-period-e

Dey N, Ashour AS, Shi F, Fong SJ, Tavares JMR (2018) Medical cyber-physical systems: a survey. J Med Syst 42:1–13

Garcia-Perez A, Cegarra-Navarro JG, Sallos MP, Martinez-Caro E, Chinnaswamy A (2023) Resilience in healthcare systems: Cyber security and digital transformation. Technovation 121:102583

Giftakis JE, Stanskaski SR, Harding WC, Gertenbach R, Bechard BL, Torgerson NA (2025) US Patent 12,280,261. Washington, DC: US Patent and Trademark Office

Groh M, Sankaranarayanan A, Singh N, Kim DY, Lippman A, Picard R (2024) Human detection of political speech deepfakes across transcripts, audio, and video. Nat Commun 15(1):7629

Harding WCB (2022) Healthcare IT leaders as stakeholders in assessing, decision-making, adopting, and sustaining of transformative technology. Doctoral dissertation, Grand Canyon University

Harding WC, Maluf-Burgman MDC (2021) US Patent 11,202,914. US Patent and Trademark Office, Washington, DC

Harding WC, Petroff N, Partridge B (2021) Wearable technology and robotics for a mobile world. In: Mobile medicine. Productivity Press, New York, pp 13–37

Hilm D, Rahim D (2019) Two-factor authentication and digital signing for an enterprise system utilizing Yubikey

IEEE/UL (2024) IEEE/UL 2933-2024 standard for clinical Internet of Things (IoT) data and device interoperability with TIPPSS—Trust, Identity, Privacy, Protection, Safety, and Security (Standard no. 2933-2024). IEEE, New York

Jarrett MP (2017) Cybersecurity—a serious patient care concern. JAMA 318(14):1319–1320

Juan H (2025) Re: Dark web statistics & trends for 2025. Blog. https://preyproject.com/blog/dark-web-statistics-trends

Keskin OF, Caramancion KM, Tatar I, Raza O, Tatar U (2021) Cyber third-party risk management: a comparison of non-intrusive risk scoring reports. Electronics 10(10):1168

Kolias C, Kambourakis G, Stavrou A, Voas J (2017) DDoS in the IoT: Mirai and other botnets. Computer 50(7):80–84

Kor M, Yitmen I, Alizadehsalehi S (2023) An investigation for integration of deep learning and digital twins towards construction 4.0. smart and sustainable. Built Environ 12(3):461–487

Lewy H (2015) Wearable technologies–future challenges for implementation in healthcare services. Healthc Technol Lett 2(1):2–5

Marjanov T, Konstantinou M, Jóźwiak M, Spagnuelo D (2023) Data security on the ground: investigating technical and legal requirements under the GDPR. Proc Priv Enhancing Technol 2023(3):405–417

Medcrypt (2025, February 13) Advancing clinical IoT with IEEE/UL 2933: a framework for trust, security, and interoperability [Video]. YouTube. https://www.youtube.com/watch?v=JQ7dh25R1Qo

Meggitt S (2018) MEDJACK attacks: the scariest part of the hospital. Tufts University

Möller DP, Haas RE (2022, October) Cybersecurity needs and benefits: the four rings model. In: International conference on information technology and applications. Springer, Singapore, pp 461–471

Myers D (2018, February 2) Even KFC's head chef does not know the secret recipe. Los Angeles Times. https://www.latimes.com/food/sns-dailymeal-1869641-eat-kfc-secret-recipe-fried-chicken-20180202-story.html

Ntalampiras S, Arsić D, Hofmann M, Andersson M, Ganchev T (2014) PROMETHEUS: heterogeneous sensor database in support of research on human behavioral patterns in unrestricted environments. SIViP 8:1211–1231

Ordonez G (2020) Cyber security in the healthcare industry. Honor scholar theses. 154, Scholarly and Creative Work from DePauw University. https://scholarship.depauw.edu/studentresearch/154

Paffel TD, Harding WC (2016) US Patent 9,467,450. US Patent and Trademark Office, Washington, DC

Proskauer (2023, February 28) Lessons learned from 2022's trade secret verdicts. PRO. https://www.proskauer.com/blog/lessons-learned-from-2022s-trade-secret-verdicts#:~:text=Cir.,to%20willful%20and%20malicious%20misappropriation

Pycroft L, Aziz TZ (2018) Security of implantable medical devices with wireless connections: the dangers of cyber-attacks. Expert Rev Med Devices 15(6):403–406

Rea-Guaman AM, San Feliu T, Calvo-Manzano JA, Sanchez-Garcia ID (2017) Comparative study of cybersecurity capability maturity models. In: International conference on software process improvement and capability determination. Springer, Cham, pp 100–113

Ryseff J, Newberry SJ, De Bruhl B (2024) The root causes of failure for artificial intelligence projects and how they can succeed. I: RAND Corporation. https://www.rand.org/pubs/research_reports/RRA2680-1.html (hämtad 2025-03-11)

Sganga N (2021a, October 7) Foreign hacking group targets hospitals, clinics with ransomware attacks, says new report. CBS News. https://www.cbsnews.com/news/cyberattacks-ransomware-hacking-hospitals-target-foreign-groups/

Sganga N (2021b, November 17) Businesses worried about cyberattacks during the holidays, report finds. CBS News. https://www.cbsnews.com/news/cyber-security-cybersecurity-ransomware-hacking-businesses-worry-holidays/

Shilko J, Riddle Z, Brooks J, Stark G, Brunner A, Goody K, Kennelly J (2021, October 7) FIN12: the prolific ransomware intrusion threat actor that has aggressively pursued healthcare targets. Manidiant. https://www.mandiant.com/resources/blog/fin12-ransomware-intrusion-actor-pursuing-healthcare-targets

Soubhagyalakshmi P, Reddy KS (2023) An efficient security analysis of bring your own device. IAES Int J Artif Intell 12(2):696

Stack L (2016) Is this the top-secret KFC recipe?. New York Times. https://www.nytimes.com/2016/08/26/dining/is-this-the-top-secret-kfc-recipe.html

Standard. SFS-EN IEC 81001-5-1:2022:en Health software and health IT systems safety, effectiveness, and security—Part 5–1: Security—Activities in the product life cycle. Finnish Standards Association. Confirmed 25.2.2022

Thielman S (2016) Yahoo hack: 1bn accounts compromised by biggest data breach in history. The Guardian 15:2016

Tu SS (2018) Three new metrics for patent examiner activity: office actions per grant ratio (OGR), office actions per disposal ratio (ODR), and grant to examiner ratio (GER). J Pat Trademark Off Soc'y 100:277

Vizziello A, Magarini M, Savazzi P, Galluccio L (2023) Intra-body communications for nervous system applications: current technologies and future directions. Comput Netw 227:109718

VMware (2022, August 8) VMware report warns of deepfake attacks and cyber extortion. https://news.vmware.com/emea. Retrieved Nov 9, 2023, from https://news.vmware.com/releases/vmware-report-warns-of-deepfake-attacks-and-cyber-extortion

Wagstaff J (2015) Medical data, cybercriminals' holy grail, now espionage target

Williams PA, Woodward AJ (2015) Cybersecurity vulnerabilities in medical devices: a complex environment and multifaceted problem. Med Devices 8:305–316

Zhang S (2015) The fake places that only exist to catch copycat mapmakers

GPSR Compliance

*The European Union's (EU) General Product Safety Regulation (GPSR) is a set of rules that requires consumer products to be safe and our obligations to ensure this.*

*If you have any concerns about our products, you can contact us on ProductSafety@springernature.com*

In case Publisher is established outside the EU, the EU authorized representative is:

Springer Nature Customer Service Center GmbH
Europaplatz 3
69115 Heidelberg, Germany

**Batch number: 09187131**

Printed by Printforce, the Netherlands